Antenna Design and Development for mm-wave Systems

Antenna Design and Development for mm-wave Systems

Amjad Iqbal and Tayeb A. Denidni

Institut national de la recherche scientifique
Montreal (Quebec)
Canada

WILEY

Published by John Wiley & Sons, Inc., Hoboken, New Jersey.

Library of Congress Cataloging-in-Publication Data has been applied for:

Print ISBN: 9781394370214
ePDF ISBN: 9781394370238
ePub ISBN: 9781394370221
oBook ISBN: 9781394370245

Cover Design: Wiley
Cover Image: Photon photo/Shutterstock

To my parents

Contents

List of Figures *xvii*
List of Tables *xxi*
About the Authors *xxiii*
Preface *xxv*
Acronyms *xxvii*
About the Companion Website *xxix*

1 Introduction *1*
1.1 Why This Book? *1*
1.2 Why Focus on mm-Wave Antennas? *2*
1.3 Intended Audience *2*
1.4 What This Book Covers *3*
1.5 Scope and Limitations *4*
1.6 Concluding Remarks *4*
1.7 Organization and Flow of the Book *5*
References *6*

2 Millimeter-wave Wireless Communication Systems *9*
2.1 Understanding the RF Spectrum: From Low Frequencies to Terahertz *11*
2.1.1 Very Low to High Frequencies (3 kHz–30 MHz) *11*
2.1.2 VHF and UHF Bands (30 MHz–3 GHz) *12*
2.1.3 Microwave Region (3–30 GHz) *12*
2.1.4 Millimeter-wave (30–300 GHz) *12*
2.1.5 THz Band (300 GHz–3 THz) *12*
2.2 Applications of mm-Wave Frequencies *13*
2.2.1 5G and Beyond Mobile Networks *13*
2.2.2 High-speed Indoor Wireless Networks *13*
2.2.3 Automotive and V2X Communications *14*

2.2.4 Wireless Backhaul and Fronthaul *14*
2.2.5 Healthcare and Medical Imaging *14*
2.2.6 Industrial Automation and Smart Factories *15*
2.2.7 Immersive Technologies: AR, VR, and XR *15*
2.2.8 Security Screening and Surveillance *15*
2.3 Impairments in mm-Wave Antennas and Communication Systems *16*
2.3.1 High Free-space Path Loss *16*
2.3.2 Poor Penetration Through Obstacles *17*
2.3.3 Sensitivity to Line-of-sight and Blockages *18*
2.3.4 Rain, Fog, and Atmospheric Absorption *18*
2.3.5 Challenges in Antenna and Circuit Design *18*
2.3.6 Beamforming and Mobility Limitations *19*
2.3.7 Thermal and Noise Issues *19*
2.4 Design Strategies for mm-Wave Antennas *20*
2.4.1 Material Considerations *20*
2.4.2 Size and Miniaturization *21*
2.4.3 Feeding Techniques *21*
2.4.4 Impedance Matching and Bandwidth Enhancement *21*
2.4.5 Array Design and Beam Steering *21*
2.4.6 Integration with Packaging *21*
2.4.7 Simulation and Fabrication Constraints *22*
2.5 Fabrication and Measurement Techniques for mm-Wave Antennas *22*
2.5.1 Fabrication Techniques *22*
2.5.2 Measurement Techniques *24*
2.5.3 Challenges and Best Practices *26*
2.6 Conclusion *26*
References *27*

3 Millimeter-wave Planar Antennas *31*
3.1 Introduction *31*
3.1.1 Why Planar Antennas for mm-Wave? *32*
3.1.2 Design Considerations and Challenges *32*
3.1.3 Applications of Planar mm-Wave Antennas *33*
3.2 Microstrip Antennas *34*
3.2.1 Basic Structure and Operation *34*
3.2.2 Advantages at mm-Wave Frequencies *35*
3.2.3 Design Considerations and Challenges *36*
3.2.4 Applications *37*
3.2.5 Design Example *37*

3.3 Aperture-coupled Antennas *40*
3.3.1 Structure and Principle of Operation *40*
3.3.2 Advantages for mm-Wave Design *40*
3.3.3 Design Challenges *41*
3.3.4 Applications *41*
3.4 SIW Antennas *41*
3.4.1 What Is SIW? *41*
3.4.2 Why SIW Is Great for mm-Wave? *42*
3.4.3 SIW Antenna Types *42*
3.4.4 Design Considerations *43*
3.4.5 Applications of SIW Antennas *43*
3.4.6 Design Example *44*
3.5 Millimeter-wave Slot Antennas *46*
3.5.1 Principle of Operation *47*
3.5.2 Slot Width and Bandwidth Considerations *48*
3.5.3 Feeding Mechanisms *48*
3.5.4 Applications and Advantages *48*
3.5.5 Design Note for Arrays *49*
3.6 Conclusion *49*
References *49*

4 Millimeter-wave Dielectric Resonator Antennas *51*
4.1 Introduction *51*
4.2 Dielectric Resonator Antennas *51*
4.3 Advantages of DRAs at mm-Wave Frequencies *52*
4.4 Dielectric Materials for DRAs *54*
4.5 Conductor and Dielectric Losses *56*
4.6 Feeding Techniques *57*
4.6.1 Coaxial Probe Feeding Technique *57*
4.6.2 Electromagnetically Coupled DRAs *58*
4.6.3 Aperture-coupled DRAs *59*
4.6.4 CPW-fed DRAs *61*
4.7 Applications of DRAs *62*
4.7.1 Wireless Communication Systems *62*
4.7.2 Automotive Radar Systems *62*
4.7.3 Wireless Sensing and Imaging *63*
4.7.4 Satellite and Point-to-point Links *63*
4.7.5 Research and Experimental Use *63*
4.8 Conclusion *63*
References *64*

5 Substrate Integrated Waveguide Based mm-Wave Antennas *67*
5.1 Introduction *67*
5.2 Single-band SIW Antennas *69*
5.2.1 Waveguide Dimensions and Effective Width *69*
5.2.2 Rules for via Placement *70*
5.2.3 Cavity-type Single-band SIW Antenna *71*
5.2.4 Radiation and Bandwidth Considerations *72*
5.2.5 Summary of Design Process *72*
5.3 Multiband SIW Antennas *72*
5.3.1 Concept of Multiband Operation *73*
5.3.2 Self-diplexing SIW Antennas *73*
5.3.3 Self-triplexing SIW Antennas *74*
5.3.4 Self-quadruplexing SIW Antennas *75*
5.3.5 Placement of Vias and Multiband Stability *76*
5.3.6 Feed Structures and Isolation Between Bands *77*
5.3.7 Advantages and Application Context *77*
5.4 SIW Antennas Arrays *77*
5.4.1 Array Configurations *78*
5.4.2 Beam Steering and Pattern Control *79*
5.4.3 Mutual Coupling Considerations *79*
5.4.4 Performance at mm-Wave Frequencies *80*
5.5 Wideband SIW Antennas *80*
5.6 Conclusion *81*
References *81*

6 Wideband Millimeter-wave Antennas *85*
6.1 Why Wideband Antennas? *85*
6.2 Bandwidth Enhancement Techniques *87*
6.3 Stacked Microstrip Antenna Techniques for Bandwidth Enhancement *88*
6.3.1 Basic Structure and Bandwidth Problem *89*
6.3.2 Principle of Bandwidth Enhancement *89*
6.3.3 Coupling Distance and Material Effects *89*
6.3.4 Feeding Techniques *90*
6.3.5 Mechanical and Manufacturing Considerations *90*
6.4 Parasitic-element Techniques for Bandwidth Enhancement *90*
6.4.1 Basic Idea of Parasitic Elements *91*
6.4.2 Spacing Between Radiator and Parasitic Element *91*
6.4.3 Different Parasitic-element Positions *92*
6.4.4 Directors and Reflectors *92*

6.4.5 Multiple Parasitic Elements *92*
6.4.6 Printed Parasitic Structures *93*
6.4.7 Aperture-type Parasitic Elements *93*
6.4.8 Bandwidth and Gain Trade-off *94*
6.4.9 Benefits for mm-Wave Systems *94*
6.4.10 Practical Design Challenges *95*
6.5 Hybrid Techniques *95*
6.5.1 Mix of Slot and Patch Antennas *96*
6.5.2 Mix of Stacked Layers and Parasitic Elements *97*
6.5.3 Mix of Different Feeds and Radiators *97*
6.5.4 Mix of Slots and Parasitic Edges *98*
6.5.5 Mix with Electromagnetic Band Gap *98*
6.5.6 Mix of Tuning Stubs and Modified Ground *98*
6.5.7 Advantages of Hybrid Techniques *98*
6.5.8 Design Challenges *99*
6.5.9 Practical Examples *99*
6.6 Conclusion *99*
References *100*

7 Millimeter-wave Planar Antenna Arrays *103*
7.1 Introduction *103*
7.1.1 What Is a Planar Antenna Array? *103*
7.1.2 Importance of mm-Wave Planar Arrays *106*
7.1.3 Design Considerations for mm-Wave Planar Arrays *107*
7.1.4 Applications of Planar Arrays *108*
7.1.5 Challenges in mm-Wave Planar Array Design *109*
7.1.6 Summary *110*
7.2 Series-fed Antenna Arrays *110*
7.2.1 Introduction *110*
7.2.2 Basic Concept of Series-fed Arrays *111*
7.2.3 Array Factor of Series-fed Arrays *112*
7.2.4 Design of Series-fed Arrays *113*
7.2.5 Transmission Line Model *114*
7.2.6 Radiation Pattern Analysis *114*
7.2.7 Advantages of Series-fed Arrays *114*
7.2.8 Limitations *114*
7.2.9 Practical Applications *115*
7.2.10 Example: Series-fed Microstrip Array *115*
7.2.11 Summary *115*
7.3 Parallel-fed Arrays *115*
7.3.1 Introduction *115*

7.3.2 Parallel Feeding Basics *116*
7.3.3 Array Factor and Radiation Pattern *118*
7.3.4 Power Division and Phase Control *118*
7.3.5 Types of Parallel Feeding Networks *119*
7.3.6 Mutual Coupling and Losses *119*
7.3.7 Applications of Parallel-fed Arrays *120*
7.3.8 Summary *120*
7.4 Combination-fed Arrays *120*
7.4.1 Introduction *120*
7.4.2 Principle of Combination Feeding *121*
7.4.3 Need for Combination-fed Arrays *121*
7.4.4 Array Factor and Mathematical Model *122*
7.4.5 Current and Phase Distribution *122*
7.4.6 Typical Combination Structures *123*
7.4.7 Design Considerations and Equations *123*
7.4.8 Beam Steering and Pattern Control *124*
7.4.9 Loss, Bandwidth, and Efficiency *125*
7.4.10 Mutual Coupling and Matching *125*
7.4.11 Comparison with Series and Parallel Arrays *125*
7.4.12 Applications *126*
7.4.13 Summary *126*
7.5 Design Considerations for Antenna Arrays *127*
7.5.1 Element Spacing *127*
7.5.2 Amplitude Distribution *127*
7.5.3 Phase Control *127*
7.5.4 Impedance Matching *128*
7.5.5 Mutual Coupling *128*
7.5.6 Feed Network Design *128*
7.5.7 Bandwidth *128*
7.5.8 Physical and Fabrication Constraints *128*
7.6 Conclusion *129*
References *129*

8 Millimeter-wave Reconfigurable Antennas Using FSS *133*
8.1 Introduction *133*
8.1.1 Challenges of Antenna Design at mm-Wave Frequencies *134*
8.1.2 Why Reconfigurable Antennas *134*
8.1.3 Role of Frequency Selective Surfaces *135*
8.1.4 Summary *137*
8.2 Fundamentals of FSS *138*
8.2.1 Definition and Basic Principle of Operation *138*

8.2.2 Transmission vs. Reflection Mode FSS *138*
8.2.3 Equivalent Circuit Models for FSS *139*
8.2.4 Types of Unit Cells *140*
8.2.5 Bandpass, Bandstop, and Polarization-dependent Designs *141*
8.2.6 Summary *142*
8.3 mm-Wave Reconfigurable Antennas *142*
8.3.1 Overview of Reconfigurability *142*
8.3.2 Frequency Reconfigurability *143*
8.3.3 Polarization Reconfigurability *143*
8.3.4 Radiation Pattern Reconfigurability *143*
8.3.5 Gain and Beam Steering Reconfigurability *144*
8.3.6 Techniques for Reconfiguration *144*
8.3.7 Challenges at mm-Wave Bands *146*
8.3.8 Summary *146*
8.4 Integration of FSS with mm-Wave Reconfigurable Antennas *146*
8.4.1 FSS as Superstrate for Gain Enhancement *147*
8.4.2 FSS as a Reconfigurable Reflector/Absorber for Beam Control *147*
8.4.3 FSS-based Polarization-reconfigurable Antennas *148*
8.4.4 FSS Metasurfaces for Multiband and Wideband Operation *148*
8.4.5 Hybrid Integration: FSS with Phased Arrays, RIS, and MIMO Antennas *148*
8.4.6 Conclusion *149*
8.5 Design Methodologies *149*
8.5.1 Unit Cell Design and Periodicity Considerations *150*
8.5.2 Simulation Approaches: CST, HFSS, and Equivalent Circuit Models *150*
8.5.3 Parametric Analysis of Unit Cell Geometry and Materials *151*
8.5.4 Biasing Techniques for Active FSS Elements *151*
8.5.5 Fabrication Challenges: Lithography, 3D Printing, and Flexible Substrates *152*
8.5.6 Summary *153*
8.6 Performance Metrics *153*
8.6.1 Reflection and Transmission Coefficient (S-parameters) *153*
8.6.2 Bandwidth and Frequency Agility *154*
8.6.3 Gain Enhancement and Beam Shaping *154*
8.6.4 Efficiency and Losses at mm-Wave *155*
8.6.5 Isolation in MIMO Configurations *155*
8.6.6 Trade-offs Between Complexity, Cost, and Performance *155*
8.6.7 Summary *156*
8.7 Applications *156*

8.7.1 5G and 6G mm-Wave Wireless Communication Systems *156*
8.7.2 Satellite and Space Communications *157*
8.7.3 mm-Wave Radar and Imaging Systems *157*
8.7.4 Wireless Power Transfer and Energy Harvesting *158*
8.7.5 Integration with RIS for Smart Environments *159*
8.7.6 Summary *159*
8.8 Challenges and Future Trends *159*
8.8.1 Fabrication and Tolerance Issues at mm-Wave Bands *159*
8.8.2 Thermal and Power Handling Constraints of Reconfigurable FSS *160*
8.8.3 Integration with CMOS and Compact Packaging *160*
8.8.4 Emergence of Tunable 2D Materials for FSS-based Reconfigurable Antennas *161*
8.8.5 AI/ML-driven Optimization for Adaptive FSS-based Antenna Systems *161*
8.8.6 Future Perspective: Convergence of FSS, RIS, and mm-Wave Antennas for 6G and THz Systems *162*
8.8.7 Summary *162*
8.8.8 Summary of the Role of FSS in Enabling Reconfigurability *162*
8.8.9 Potential for Next-generation Wireless and Sensing Systems *163*
8.8.10 Open Research Directions *164*
8.8.11 Summary Table of Key Points *165*
8.9 Conclusion *165*
References *166*

9 Orbital Angular Momentum Antennas *169*
9.1 Introduction *169*
9.2 Fundamentals of OAM *171*
9.2.1 Angular Momentum in Electromagnetics *171*
9.2.2 Orthogonality of OAM Modes *172*
9.3 OAM Generation *172*
9.3.1 Spiral Phase Plates *172*
9.3.2 Uniform Circular Arrays *173*
9.3.3 Phased Arrays with Progressive Phase Shift *173*
9.3.4 Metasurfaces *173*
9.3.5 Dielectric Resonator Modes *174*
9.3.6 Array Factor for OAM Modes *174*
9.4 Challenges in OAM Antennas *174*
9.4.1 Beam Divergence *174*
9.4.2 Mode Purity and Cross-talk *175*

9.4.3 Multipath Sensitivity *176*
9.4.4 Receiver Architectures *176*
9.5 Applications of OAM Antennas *176*
9.5.1 Point-to-point High-capacity Links *176*
9.5.2 Secure Communications *177*
9.5.3 Chip-to-chip Wireless Links *177*
9.5.4 Radar and Imaging Applications *177*
9.6 Future Outlook *177*
9.7 Design Example *178*
9.7.1 Literature Review *178*
9.7.2 Unit Cell Characteristics *180*
9.7.3 Mixed-OAM generating TA Design *181*
9.7.4 Experimental Verification *183*
9.7.5 Discussion *186*
9.7.6 Summary *187*
9.8 Conclusion *188*
References *188*

10 Conclusion *193*

Index *195*

List of Figures

2.1 RF spectrum from low frequencies to terahertz. *11*
2.2 Key impairments in mm-wave communication systems. *17*
2.3 Summary of key design strategies for mm-wave antennas. *20*
2.4 Overview of key fabrication techniques for mm-wave antennas. *23*
2.5 Measurement techniques for mm-wave antennas. *25*
3.1 CPW-fed microstrip planar antenna. *32*
3.2 CPW-fed planar patch antenna. *33*
3.3 SIW-fed 5G antenna. *33*
3.4 Rectangular microstrip patch antenna. *34*
3.5 Main design considerations and challenges for millimeter-wave microstrip antennas. *36*
3.6 Geometry of the integrated microwave and mm-wave antenna ($A = 16$, $L = 21$, $h = 1.6$, $g_m = 3$, $w_s = 2$, $g_w = 4.1$, $w_p = 13$, $p_w = 5$, $l_1 = 10.5$, $w_1 = 13.7$, $l_2 = 8.2$, $w_2 = 9.2$, $l_3 = 5.9$, $w_3 = 4.8$, $l_4 = 3.5$, $w_4 = 5$, $l_5 = 2$, $s = 1$, $w = 1$, $w_o = 3.2$, $w_a = 14.2$, $w_b = 9.2$, $l_a = 10.5$, and $l_b = 8.2$ [unit = mm]). *38*
3.7 Design evolution. *38*
3.8 S_{11} of the design evolution. *39*
3.9 SIW-based microwave and mm-wave antenna configuration. *44*
3.10 Reflection coefficient of the antenna. *45*
3.11 Mutual coupling between the ports of the antenna. *45*
3.12 Reflection coefficient by varying parameter "L". *46*
3.13 Reflection coefficient by varying the number of vias. *46*
3.14 Configuration of mm-wave SIW-based slot antenna. *47*
4.1 Illustration of dominant resonant modes in a cylindrical DRA. *52*
4.2 Summary of key material considerations for mm-wave dielectric resonator antennas. *56*
4.3 Qualitative comparison of conductor and dielectric losses in common mm-wave antenna technologies. *57*

4.4 Illustration of coaxial probe feeding technique for circular DRA. 58
4.5 Illustration of electromagnetically coupled circular DRA. 59
4.6 Illustration of aperture-coupled circular DRA. 60
4.7 Illustration of CPW-fed circular DRA. 61
4.8 Visual representation of different applications of DRAs. 62
5.1 Example of two-port SIW structure. 68
5.2 SIW-DRA-based single-band antenna. 70
5.3 SIW-based self-diplexing dual-band antenna. 74
5.4 SIW-based self-triplexing triple-band antenna. 75
5.5 SIW-based microwave and mm-wave self-quadruplexing antenna configuration. 76
5.6 SIW-based mm-wave antenna array configuration. 78
6.1 An example of a two-layer microstrip patch antenna excited using the probe feeding technique. 88
6.2 An example of a microstrip patch antenna coupled with two parasitic elements. The main patch is excited by the probe feed method. 91
6.3 An example of a microstrip patch antenna coupled with multiple parasitic elements. The main resonator (patch) is excited by the probe feed method. 93
6.4 An example of a microstrip patch antenna coupled with two parasitic elements and also with a stacked patch. The main patch is excited by the probe feed method. 96
7.1 An example of 2 × 2 planar patch array. 104
7.2 S-parameters of the 2 × 2 planar patch array. 105
7.3 Radiation pattern of the 2 × 2 planar patch array. 105
7.4 Key design considerations for millimeter-wave planar antenna arrays. 107
7.5 Series-fed planar array. 111
7.6 S-parameters of the series-fed four-element planar array. 111
7.7 Parallel-fed planar array. 116
7.8 S-parameters of the parallel-fed planar array. 117
7.9 Radiation pattern of the parallel-fed planar array. 117
8.1 Reconfigurable FSS geometry. 136
8.2 FSS as superstrate geometry. 136
8.3 (a) Geometry of unit cell of the absorber; (b) layout of the resonant structure of the absorber. 137
8.4 FSS 7 × 7 array and its unit cell. 137
8.5 FSS geometry: (a) 3D view of the panel, (b) side view of the FSS panel, and (c) top view of the unit cell. 139

8.6 (a) Surface current distribution; (b) equivalent circuit of the active FSS. *140*
8.7 Dipole-based FSS. *141*
8.8 Geometry of the fractal FSS. *141*
8.9 (a) The geometry of tunable FSS; (b) equivalent circuit model. *145*
8.10 (a) The geometry of FSS; (b) circuit model. *145*
8.11 Summary of key aspects. *165*
9.1 Key methods for generating OAM modes. *173*
9.2 Key challenges in OAM antennas. *175*
9.3 Huygens' metasurface: (a) top view, (b) bottom view, (c) simulation environment, and (d) unit cell parameters and their values. *180*
9.4 Unit cell results for different l values: (a) phase response and (b) amplitude response. *181*
9.5 Schematic of 6×6 elements of TA. *182*
9.6 Fabricated prototype: (a) top view, (b) bottom view, and (c) near-field measurement setup. *184*
9.7 Results at 28 GHz: (a) X-polarization and (b) Y-polarization. *184*
9.8 X- and Y-polarized measured patterns at 28 GHz. *185*
9.9 Mode purity results: (a) $r = 4\lambda$, (b) $r = 8\lambda$, and (c) $r = 12\lambda$. *185*

List of Table

7.1 Comparison of series-fed, parallel-fed, and combination-fed arrays. *126*

About the Authors

Amjad Iqbal received the B.S. degree in Electrical Engineering from COMSATS University Islamabad, Pakistan, in 2016, the M.S. degree in Electrical Engineering from CECOS University of IT and Emerging Sciences, Pakistan, in 2018, and the Ph.D. degree in Engineering from Multimedia University, Malaysia, in 2021.

From 2016 to 2018, he worked as a Laboratory Engineer in the Department of Electrical Engineering at CECOS University, Pakistan. He is currently a Postdoctoral Fellow at the Institut National de la Recherche Scientifique, Canada. His research interests include printed and flexible antennas, implantable antennas, MIMO systems, dielectric resonator antennas, wireless power transfer, and the synthesis of microwave and millimeter-wave components.

Tayeb A. Denidni received the M.Sc. and Ph.D. degrees in Electrical Engineering from Université Laval, Quebec City, QC, Canada, in 1990 and 1994, respectively. From 1994 to 2000, he was a Professor with the Department of Engineering at Université du Québec à Rimouski (UQAR), Rimouski, QC, Canada, where he founded the Telecommunications Laboratory.

Since 2000, he has been with the Institut National de la Recherche Scientifique (INRS), University of Quebec, Montreal, QC, Canada. He established the RF Laboratory at INRS–Énergie, Matériaux et Télécommunications (INRS-EMT), Montreal.

He has extensive experience in antenna design and has served as a Principal Investigator on numerous research projects funded by NSERC, FCI, and various industry partners. His current research interests include reconfigurable antennas based on electromagnetic bandgap and frequency-selective surface structures, dielectric resonator and metamaterial antennas, adaptive and switched-beam arrays, ultra-wideband antennas, and microwave systems for wireless communications.

Preface

Wireless technology has changed the way we live, work, and communicate. Over the past few decades, the demand for faster data, higher capacity, and more reliable connections has continued to grow. This demand is now pushing communication systems beyond traditional frequency bands and into the millimeter-wave spectrum. While this shift opens new opportunities, it also introduces new challenges, especially in antenna design.

This book was written to address that gap. Many existing resources either focus heavily on theory or present isolated design examples without practical context. In contrast, this book brings both together. It explains the fundamental concepts of millimeter-wave antennas while also showing how these designs are implemented, tested, and integrated into real systems.

The idea behind this book developed through years of research, teaching, and collaboration. Working closely with students, researchers, and industry partners made it clear that there is a need for a structured and practical guide that connects electromagnetic theory with modern wireless applications. This book reflects that experience.

The content is organized to support a gradual learning process. It starts with basic concepts and system-level understanding, then moves toward specific antenna types, design techniques, and advanced topics such as reconfigurable structures and orbital angular momentum. Each chapter is written to be readable on its own while still contributing to a complete picture of the field.

This book is intended for graduate students, researchers, and practicing engineers working in antennas, electromagnetics, and wireless systems. It can be used both as a learning resource and as a reference during design and research work. The goal is not only to explain how antennas work, but also to help you think about design choices, trade-offs, and real-world constraints.

Millimeter-wave technology will play a key role in future communication systems, including fifth-generation, sixth-generation, automotive radar, satellite links, and emerging sensing applications. Antennas remain a central part of these systems. Their performance often defines the overall system capability. For this reason, understanding antenna behavior at these frequencies is important.

This book does not try to cover everything. Instead, it focuses on what matters most for antenna engineers: practical design approaches, clear explanations, and relevant applications. Some topics are simplified where needed, while others are discussed in more detail to support advanced work.

We hope this book helps you build a clear understanding of millimeter-wave antennas and supports your work, whether in research, teaching, or engineering practice.

Amjad Iqbal
Tayeb A. Denidni

Acronyms

5G	Fifth-generation Wireless
6G	Sixth-generation Wireless
AoA	Angle of Arrival
AoD	Angle of Departure
BT	Bluetooth
CPW	Coplanar Waveguide
CST	Computer Simulation Technology
DRA	Dielectric Resonator Antenna
EIRP	Effective Isotropic Radiated Power
EM	Electromagnetic
FBR	Front-to-back Ratio
FSS	Frequency Selective Surface
HFSS	High-frequency Structure Simulator
LNA	Low-noise Amplifier
LOS	Line-of-sight
LTE	Long Term Evolution
MIMO	Multiple-input Multiple-output
mm-Wave	Millimeter Wave
NLOS	Non-line-of-sight
NR	New Radio
OAM	Orbital Angular Momentum
PA	Power Amplifier
PCB	Printed Circuit Board
RF	Radio Frequency
RIS	Reconfigurable Intelligent Surface
SAR	Specific Absorption Rate
SIW	Substrate Integrated Waveguide
VSWR	Voltage Standing Wave Ratio
Wi-Fi	Wireless Fidelity

About the Companion Website

This book is accompanied by a companion website.

https://wiley.com/go/iqbalmmwave

To access the website materials, click on the link under Related Resources on the book's page on wiley.com. You can also find the website by searching www.wiley.com for the ISBN 9781394370214.

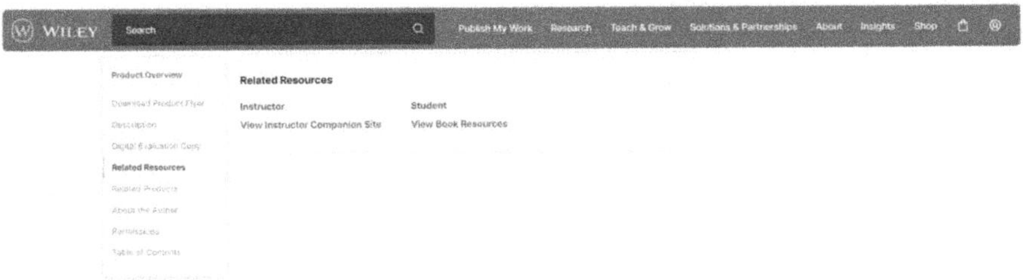

This website includes:

- MCQs in PDF
- PowerPoint (PPT) slides

1

Introduction

1.1 Why This Book?

Wireless communication systems have evolved dramatically over the past few decades: from large, clunky mobile phones that only supported basic calls and texts to today's compact smartphones that stream high-definition video, support real-time global communication, and run powerful applications (Niu et al., 2015). At the heart of this evolution lies a constant push for higher speeds, larger data capacities, and more efficient spectrum utilization. Currently, the sub-6 GHz frequency spectrum is overcrowded, and is not enough to cover the current demands, and communication with large spectrum utilization is required (Rappaport et al., 2015). In fact, future-generation wireless communication systems need a larger bandwidth for high-speed wireless communication, which is impossible to use in sub-6 GHz frequency spectrum. In contrast, the millimeter-wave (mm-wave) frequency band has huge available spectrum which can be used without any restriction (Wang et al., 2018). Therefore, substantial work has been put forward to develop new wireless communication systems operating the mm-wave frequency band.

In this perspective, antennas act as the ears and eyes of modern communication systems, making it possible for devices and systems to exchange information wirelessly (Balanis, 2016). Therefore, this book focuses on mm-wave antennas and how they contribute to the advancement of fifth-generation (5G) and sixth-generation (6G) wireless communication. Although there are many books available on antennas and wireless systems, only a few cover mm-wave antennas in the context of emerging technologies. Some of these works deal mainly with theoretical concepts, while others concentrate on practical design. However, very few strike a balance between theory, practical applications, and system-level integration. This book aims to fill that gap by offering both a strong theoretical foundation and practical guidance. It also provides insights into how antennas perform when integrated into real systems. In addition, it discusses the practical challenges engineers face when designing and implementing mm-wave antennas for next-generation technologies.

1.2 Why Focus on mm-Wave Antennas?

The mm-wave frequencies refer to the range above traditional microwave frequencies. Typically, they span from 30 to 300 GHz, corresponding to wavelengths between 10 and 1 mm. These frequencies offer several advantages over conventional microwave frequencies (Sun et al., 2014). Antennas at these frequencies can be significantly smaller, with a tiny footprint, allowing for high-density integration and compact arrays. Additionally, the wider bandwidths available at mm-wave enable ultra-high data rates, which are crucial for applications like real-time virtual reality, 8K streaming, autonomous vehicles, and massive internet-of-things (IoT) deployments.

However, the mm-wave band is not without its difficulties. Signals in this range suffer from increased free-space path loss, are more susceptible to obstruction from objects such as buildings, walls, or even human bodies, and experience greater attenuation due to rain and atmospheric absorption (Busari et al., 2018). To overcome these issues, mm-wave antenna systems must be highly efficient, directional, and, in many cases, reconfigurable to ensure robust and reliable communication links.

The aim of this book is to equip readers with both the theoretical background and the practical tools needed to overcome these challenges, and to view mm-wave antennas not as standalone components, but as integrated parts of a larger wireless system.

1.3 Intended Audience

This book is tailored for a broad audience, including:

- **Graduate students** pursuing studies in antennas, electromagnetics, wireless communications, or microwave engineering.
- **Researchers** working in academia or industry who need a comprehensive reference on mm-wave antenna design.
- **Professional engineers** involved in the development of next-generation communication systems, radar, automotive, biomedical, and satellite technologies.

For newcomers, the early chapters offer foundational knowledge that does not assume advanced expertise in electromagnetics or antenna design. For experienced practitioners, the book provides advanced design examples, real-world case studies, and discussions on simulation and measurement techniques that are directly applicable to ongoing projects.

1.4 What This Book Covers

This book provides a comprehensive and application-focused exploration of mm-wave antennas designed for modern wireless communication systems. It spans from foundational principles to advanced antenna architectures and cutting-edge innovations, offering readers both depth and breadth in the field of high-frequency antenna design.

At the core, the book begins by introducing the reader to the context and importance of mm-wave wireless systems. It explains how the growing demand for high-data rates, massive device connectivity, and low-latency services in technologies like 5G and beyond could boost mm-wave communication to the forefront. The book highlights the unique challenges posed by the mm-wave spectrum, including signal attenuation, material losses, and hardware limitations, while also offering strategies for mitigating these issues through careful antenna and system design.

Subsequent chapters cover a wide range of antenna structures tailored for mm-wave operation. Planar antennas, which include microstrip, aperture-coupled, slot, and substrate integrated waveguide (SIW) configurations, are thoroughly discussed. These low-profile and compact designs are especially relevant for integrated and wearable technologies.

The book then delves into dielectric resonator antennas (DRAs), presenting them as efficient and high-performing alternatives to conventional designs. Different coupling methods, such as coaxial probe, aperture coupling, and coplanar waveguide (CPW), feeding are explained with practical examples and applications.

Building on the concept of SIWs, the book explores SIW-based antennas in detail. Readers learn how SIW technology offers a low-loss, cost-effective solution for both narrowband and broadband mm-wave systems, and how it enables scalable antenna arrays suitable for beamforming and multiple-input multiple-output systems.

Recognizing the growing need for bandwidth, a dedicated chapter addresses wideband mm-wave antenna technologies. It presents various bandwidth enhancement techniques including stacked patches, parasitic elements, and hybrid configurations, making the designs more resilient to frequency-selective fading and useful in multistandard wireless platforms.

The book also emphasizes the importance of antenna arrays at mm-wave frequencies, focusing on array configurations, feeding networks, and critical design trade-offs. This is followed by an exploration of reconfigurable antennas using frequency selective surfaces (FSS), which are increasingly being used to enable dynamic control over antenna parameters such as frequency, polarization, and radiation pattern.

The final technical chapter introduces the concept of orbital angular momentum (OAM) in mm-wave antenna systems. OAM-based antennas represent a frontier in wireless communications, offering new ways to increase spectral efficiency and secure data transmission through the use of twisted EM waves.

Each chapter not only discusses theoretical aspects but also covers design practices, material selection, fabrication techniques, and real-world applications, providing engineers, researchers, and students with the tools and insights needed to implement high-performance mm-wave antennas for next-generation wireless systems.

This book serves as both a solid academic resource and a practical design guide, bridging the gap between microwave theory and modern wireless engineering.

1.5 Scope and Limitations

While this book is comprehensive in its treatment of mm-wave antennas, it does not attempt to cover the entire wireless communication stack. Topics such as signal processing algorithms, modulation schemes, or network protocols are only briefly discussed when they directly relate to antenna system performance.

Additionally, the focus is primarily on the frequency range of 24–100 GHz, which encompasses most current commercial and experimental mm-wave systems. Frequencies beyond 100 GHz, including the emerging terahertz (THz) domain, are discussed primarily from a future outlook perspective.

1.6 Concluding Remarks

Millimeter-wave technology is more than just an academic curiosity; it's a cornerstone of next-generation communication systems. Designing efficient, compact, and reliable mm-wave antennas opens the door to innovations across countless industries: from real-time 6G networks to remote healthcare and intelligent transportation systems.

This book is your guide through that journey. Whether you're taking your first steps into mm-wave antenna design or aiming to push the boundaries of what is possible in the field, the pages ahead are designed to help you think critically, design effectively, and innovate confidently.

1.7 Organization and Flow of the Book

This book is structured to offer a clear and gradual pathway into the world of mm-wave antenna design for wireless communication systems. The chapters in this book are organized in a way that makes it easy to follow: from the basics to more advanced topics and recent developments. Each chapter can be read on its own, but together they build a complete and connected understanding of the subject.

Chapter 2 lays the groundwork by introducing readers to mm-wave wireless communication systems. This chapter walks you through the basics, like what the radio-frequency spectrum is, where mm-wave technology is used, and the kinds of signal issues you might run into at these high frequencies. It also looks at how these systems are designed, along with the tools and techniques used to build and test them. The goal is to give you a clear picture of why mm-wave antennas need special attention and what makes designing them particularly challenging.

Chapter 3 takes a more hands-on approach, focusing on planar antennas: an important type of antenna that is especially useful in compact, integrated systems. This chapter explores various types of planar antennas, including microstrip, aperture-coupled, slot, and SIW antennas. It provides detailed design considerations, typical performance characteristics, and insight into their relevance for mm-wave applications.

Chapter 4 delves into DRAs, a powerful alternative to conventional planar antennas. The chapter covers the theoretical basis, material choices, and feeding mechanisms used in DRA designs. Special attention is given to various coupling techniques such as coaxial probe, EM, aperture, and CPW-fed configurations. Readers also gain insight into DRA performance, losses, and practical implementation in real-world systems.

Chapter 5 builds on the SIW concept introduced earlier by focusing exclusively on SIW-based mm-wave antennas. This chapter examines a range of SIW antenna designs, from single-band and multiband antennas to wideband and array configurations. It addresses their advantages in terms of low-loss performance, ease of integration, and suitability for mass production, making them ideal candidates for modern communication systems.

Chapter 6 is dedicated to wideband mm-wave antennas: an increasingly important area given the bandwidth demands of 5G and 6G networks. The chapter starts by discussing bandwidth limitations at mm-wave frequencies and then introduces several techniques for bandwidth enhancement. These include stacked microstrip structures, parasitic elements, and hybrid approaches that combine multiple bandwidth-widening strategies.

Chapter 7 shifts focus to antenna arrays, which are essential for high-gain, beam-steerable communication systems. This chapter breaks down different feeding architectures, including series-fed, parallel-fed, and hybrid-fed arrays. It also highlights key design considerations and trade-offs, making it a critical reference for engineers designing high-performance array systems for mm-wave bands.

Chapter 8 introduces FSS and their role in enabling reconfigurable antennas. The chapter discusses different FSS geometries, their electromagnetic behavior, and how they can be integrated into antenna systems for dynamic performance tuning. The use cases and their practical uses are explained in the last section of this chapter.

Chapter 9 concludes the technical content by exploring one of the most cutting-edge topics in antenna research: orbital angular momentum. This chapter examines various antenna structures capable of generating OAM modes, discusses feeding techniques, and outlines the design considerations for implementing OAM in mm-wave systems. OAM has real potential to boost both channel capacity and communication security, which makes it an especially exciting area to explore.

This book is meant to be both thorough and useful. Whether you are a graduate student just getting started, a researcher diving deeper into specific topics, or an engineer working on real-world mm-wave antenna systems, you will find the content flexible and easy to navigate. You can read it from start to finish or simply skip to the chapters that are most relevant to your work.

Each chapter ends with a concise conclusion and curated references to guide further exploration. Throughout the book, theory is balanced with design insights and real-world applications, reflecting the evolving landscape of mm-wave wireless communication and antenna technologies.

References

Constantine A. Balanis. *Antenna Theory: Analysis and Design*. John Wiley & Sons, 2016.

Sherif Adeshina Busari, Kazi Mohammed Saidul Huq, Shahid Mumtaz, Linglong Dai, and Jonathan Rodriguez. Millimeter-wave massive mimo communication for future wireless systems: A survey. *IEEE Communications Surveys & Tutorials*, 20(2):836–869, 2018. https://doi.org/10.1109/COMST.2017.2787460.

Yong Niu, Yong Li, Depeng Jin, Li Su, and Athanasios V Vasilakos. A survey of millimeter wave communications (mmwave) for 5G: Opportunities and challenges. *Wireless networks*, 21:2657–2676, 2015.

Theodore S. Rappaport, George R. MacCartney, Mathew K. Samimi, and Shu Sun. Wideband millimeter-wave propagation measurements and channel models for future wireless communication system design. *IEEE Transactions on Communications*, 63(9):3029–3056, 2015. https://doi.org/10.1109/TCOMM.2015.2434384.

Shu Sun, Theodore S. Rappaport, Robert W. Heath, Andrew Nix, and Sundeep Rangan. Mimo for millimeter-wave wireless communications: Beamforming, spatial multiplexing, or both? *IEEE Communications Magazine*, 52(12):110–121, 2014. https://doi.org/10.1109/MCOM.2014.6979962.

Xiong Wang, Linghe Kong, Fanxin Kong, Fudong Qiu, Mingyu Xia, Shlomi Arnon, and Guihai Chen. Millimeter wave communication: A comprehensive survey. *IEEE Communications Surveys & Tutorials*, 20(3):1616–1653, 2018. https://doi.org/10.1109/COMST.2018.2844322.

2

Millimeter-wave Wireless Communication Systems

In recent years, the use of wireless communication systems has grown rapidly, faster than at any point in history. This trend shows no signs of slowing down. As we move toward the next generation of wireless technologies, the demand for high-speed, high-capacity, and low-latency communication continues to rise. These new technologies will require significantly higher channel capacities and broader spectrum availability to meet the performance expectations of future services and devices.

Traditional wireless communication systems, including amplitude modulation (AM), frequency modulation (FM), wireless fidelity (Wi-Fi), wireless local area networks (WLANs), satellite communications, cellular networks, television broadcasting, and Global Positioning System (GPS): mostly operate in the sub-6 GHz frequency band. These systems typically use fractional bandwidths in the range of 300 MHz–3 GHz (Dutta et al., 2016; Naqvi and Lim, 2018; Qingling and Li, 2006; Wang et al., 2018). However, the sub-6 GHz spectrum is now heavily congested and is not enough to support new services and products. In addition, with so many users and services operating in this band, there is little room left to support high-bandwidth applications efficiently. To overcome these limitations, the wireless industry is shifting its focus toward the millimeter-wave (mm-wave) frequency range. This portion of the spectrum, typically defined as 30–300 GHz, offers vast unused bandwidth that is well-suited for high-speed data transmission. Unlike sub-6 GHz bands, mm-wave frequencies are not yet heavily regulated or crowded, giving designers and engineers more freedom to innovate. There are several key advantages to using mm-wave frequencies for wireless communication (Doan et al., 2004; Mumtaz et al., 2016):

1. **Larger Available Bandwidth:** One of the most significant benefits of mm-wave frequencies is the large available spectrum. This abundance of bandwidth enables data rates that are far higher than what is possible with current systems. It makes technologies such as fifth-generation (5G),

augmented reality (AR), and ultra-high-definition (UHD) streaming feasible. The relationship between channel capacity and bandwidth can be determined using Shannon's theorem:

$$C = B \log_2 \left(1 + \frac{S}{N}\right), \tag{2.1}$$

where C is the channel capacity, B is the bandwidth, and S/N is the signal-to-noise ratio (SNR).

2. **Smaller Antenna Size:** The antenna size is inversely proportional to the operating frequency. This means that as frequency increases, the size of the antenna decreases. Therefore, operating at mm-wave frequencies allows the development of compact antennas with a small footprint, making it easier to integrate multiple antennas and RF components into small, portable devices.
3. **Massive Multiple-input Multiple-output (MIMO) and Beamforming:** Because mm-wave antennas are physically small, many of them can be placed within a compact array. This makes it practical to implement massive MIMO systems and beamforming techniques. These methods improve the signal directionality and reduce interference, which is especially important for high-density user environments.

However, along with these benefits come several challenges. One of the primary drawbacks of mm-wave communication is the high propagation loss. These signals are more easily absorbed or blocked by obstacles such as buildings, trees, or even the human body. The short wavelengths, ranging from 1 to 10 mm, also make mm-wave signals more sensitive to environmental conditions like fog, rain, and moisture. As a result, signal attenuation can be quite severe compared to traditional microwave frequencies.

Despite these limitations, mm-wave communication is still highly effective for short-range, high-data-rate links, particularly in the 28 and 38 GHz bands, where many commercial and research systems are now being deployed (Chen et al., 2016; Qingling and Li, 2006). These frequencies perform well in line-of-sight (LOS) conditions but degrade rapidly when physical obstructions are present (Rappaport et al., 2017; Reniers et al., 2020; Wang et al., 2015). The unique propagation characteristics at mm-wave frequencies make antenna design more critical than ever. While higher frequencies allow greater antenna integration and miniaturization, they also introduce several challenges. For example, fabrication tolerances become tighter; even small imperfections can have a significant impact on performance. The choice of substrate materials becomes more important as well, with higher demands on dielectric stability and loss characteristics.

2.1 Understanding the RF Spectrum: From Low Frequencies to Terahertz

The RF spectrum encompasses a wide range of electromagnetic waves used for wireless communication. It begins at a few kilohertz and extends up to terahertz (THz) frequencies. Each band in this spectrum has distinct characteristics that make it suitable for specific applications. This section walks through the RF spectrum from the lower frequencies to the THz range, highlighting their uses, limitations, and significance. RF spectrum from low frequencies to THz and its typical applications is illustrated in Figure 2.1.

2.1.1 Very Low to High Frequencies (3 kHz–30 MHz)

At the lower end of the spectrum, RF signals have long wavelengths that enable them to travel great distances and diffract around terrain and buildings:

- **Very Low Frequency (VLF):** 3–30 kHz—used in submarine communication due to its deep penetration through water and Earth.
- **Low Frequency (LF):** 30–300 kHz—historically used for navigation; supports long-range communication.
- **Medium Frequency (MF):** 300–3000 kHz—primarily for AM radio broadcasting.
- **High Frequency (HF):** 3–30 MHz—used in shortwave radio and international broadcasting due to ionospheric reflection.

Despite their range, these frequencies have limited data-handling capacity and are significantly affected by atmospheric noise.

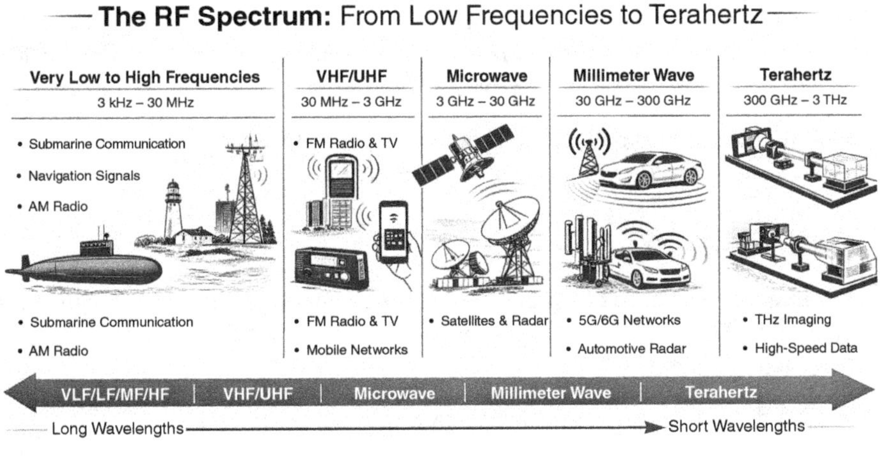

Figure 2.1 RF spectrum from low frequencies to terahertz.

2.1.2 VHF and UHF Bands (30 MHz–3 GHz)

As we move higher in frequency, the wavelengths become shorter, allowing better data rates and improved audio/video quality:

- **Very High Frequency (VHF):** 30–300 MHz—this band supports FM radio transmission, television broadcast services, and various aviation communication systems.
- **Ultra High Frequency (UHF):** 300 MHz–3 GHz—this band supports a wide range of technologies, including mobile networks, Wi-Fi connectivity, GPS positioning, and Bluetooth communication.

2.1.3 Microwave Region (3–30 GHz)

Often referred to as the microwave band, this frequency range supports high data rates and is widely used in modern communication and radar systems:

- Applications include satellite communications, radar, backhaul microwave links, and Wi-Fi at 5 GHz.
- Antennas are typically small and highly directional.
- Atmospheric absorption becomes more significant, particularly in the presence of rain and foliage.

2.1.4 Millimeter-wave (30–300 GHz)

The mm-wave frequencies have wavelengths between 1 and 10 mm and are crucial for high-speed wireless systems:

- Ideal for 5G/6G mobile networks, automotive radar, and point-to-point (P2P) communication.
- Enable high-data throughput and dense antenna arrays (MIMO systems).
- Limitations include short range and high atmospheric attenuation, particularly due to water vapor and oxygen absorption.

To overcome these challenges, technologies such as beamforming, phased array antennas, and intelligent reflecting surfaces are employed.

2.1.5 THz Band (300 GHz–3 THz)

This frequency band, still largely in the experimental stage, holds significant potential for future ultra-high-speed wireless communication systems.

- Useful for imaging, spectroscopy, and short-range data links.
- Offers high security and massive bandwidth.
- Challenges include high free-space path loss (FSPL), limited source/detector technology, and atmospheric absorption.

The RF spectrum spans from long-wavelength VLF to high-frequency THz waves. Each range has distinct properties and is optimized for specific tasks, from global broadcasting to future gigabit wireless networks. Understanding these bands is essential for developing effective antennas, communication systems, and RF technologies.

2.2 Applications of mm-Wave Frequencies

The rapid evolution of wireless technologies has consistently demanded more bandwidth, faster data transmission, and lower latency. Traditional sub-6 GHz frequency bands, while effective for decades, are quickly becoming saturated due to the proliferation of smart devices and data-intensive applications. In this context, mm-wave communication, operating in the 30–300 GHz frequency range, has emerged as a transformative technology. Initially explored for defense and satellite systems, mm-wave communication has now found a broad array of applications across consumer, industrial, and scientific domains (Chittimoju and Yalavarthi, 2021; Lopez et al., 2019; Sheen et al., 2007). This section outlines some of the most promising and emerging use cases that are shaping the future of mm-wave systems.

2.2.1 5G and Beyond Mobile Networks

One of the most prominent applications of mm-wave communication is in 5G cellular networks, and it is anticipated to be pivotal in the development of 6G technologies (Zada et al., 2021). mm-Wave bands offer significantly wider bandwidths than legacy systems, allowing data rates to reach several gigabits per second. This capability is crucial for applications like UHD video streaming, AR, and virtual reality (VR), which are data-intensive and require near-instantaneous transmission.

In urban environments, mm-wave small cells can be deployed on lamp posts, buildings, and other urban furniture to provide localized, high-capacity coverage. While propagation challenges like signal blockage and atmospheric absorption exist, advancements in beamforming, MIMO, and intelligent reflecting surfaces (IRS) are being used to enhance signal reliability and coverage.

2.2.2 High-speed Indoor Wireless Networks

The mm-wave technology is also making significant strides in indoor environments (Manabe et al., 1995). In office spaces, homes, and public venues like airports or stadiums, mm-wave communication can support ultra-fast WLANs. Technologies like IEEE 802.11ad (WiGig), which operates at 60 GHz, enable data transfer rates of up to 7 Gbps, making it suitable for wirelessly docking laptops, transferring large files, or streaming 4K/8K content without buffering.

Moreover, mm-wave's limited penetration through walls becomes an advantage in this context: it minimizes interference between adjacent rooms or offices, thereby allowing better frequency reuse and higher overall network capacity in dense deployments.

2.2.3 Automotive and V2X Communications

Another emerging field for mm-wave is the automotive industry, particularly in vehicle-to-everything (V2X) communication (Choi et al., 2016). As vehicles become more autonomous and connected, the need for high-speed, low-latency communication grows. Indeed, mm-wave offers the bandwidth necessary for vehicles to share real-time data with other cars (V2V), infrastructure (V2I), and pedestrians (V2P).

Use cases include high-resolution map updates, cooperative collision avoidance, and streaming sensor data (such as LIDAR and radar) between vehicles. While traditional automotive radars operate in the 24 and 77 GHz bands, future systems are expected to integrate mm-wave communication modules for both data transmission and object detection, enabling seamless coordination in connected and autonomous vehicle (CAV) environments.

2.2.4 Wireless Backhaul and Fronthaul

With the deployment of dense 5G small cells, there is a growing demand for high-capacity wireless backhaul and fronthaul links (Tezergil and Onur, 2022). Fiber optics is ideal but expensive and logistically challenging in some areas. mm-Wave communication provides a compelling alternative, offering fiber-like speeds without the need for extensive infrastructure.

Microwave and mm-wave P2P links operating at 70/80 GHz (E-band) are already being deployed to connect base stations in urban areas. These links offer multi-gigabit speeds and can be quickly installed, making them ideal for rapidly expanding mobile networks. Moreover, mm-wave backhaul is crucial in areas where physical cabling is impractical—such as hilly terrains or developing regions.

2.2.5 Healthcare and Medical Imaging

In the healthcare sector, mm-wave is finding niche but growing applications (Topfer and Oberhammer, 2015). One promising area is noninvasive imaging. mm-Wave frequencies can penetrate fabrics and some biological tissues, making them suitable for applications such as skin cancer detection, breast tumor imaging, and real-time monitoring of vital signs.

For instance, mm-wave-based sensors can detect minute physiological movements, such as breathing and heartbeats, from a distance. This enables continuous patient monitoring in hospitals or at home without the need for physical contact, reducing infection risks and improving patient comfort. Furthermore, the high spatial resolution afforded by mm-wave makes it a candidate for future biomedical imaging systems.

2.2.6 Industrial Automation and Smart Factories

Industry 4.0 envisions fully interconnected, smart manufacturing environments, where machines, sensors, and control systems communicate wirelessly. The mm-wave is particularly attractive in these environments due to its ability to deliver low-latency, high-throughput links that support real-time control and automation (Cheffena, 2016).

Applications include wireless machine-to-machine (M2M) communication, remote control of robotic arms, and high-resolution video feeds for quality assurance. Since many industrial settings are contained within metallic enclosures or clearly defined zones, mm-wave systems can be engineered to optimize performance in these challenging environments, especially when combined with beam steering and reflective surfaces.

2.2.7 Immersive Technologies: AR, VR, and XR

AR, VR, and extended reality (XR) demand exceptionally high data rates and minimal latency to deliver seamless user experiences. Any lag or drop in resolution can break immersion and cause discomfort, especially in VR (Shvetsov and Hamood Alsamhi, 2024; Struye et al., 2024).

Millimeter-wave technology can wirelessly transmit the massive volumes of data required by these applications, particularly when offloading computational tasks to edge devices or cloud servers. This opens up possibilities for lightweight, wireless headsets capable of real-time rendering and interaction—making mm-wave communication a foundational enabler of the next generation of immersive technologies.

2.2.8 Security Screening and Surveillance

Millimeter-wave systems are increasingly being used for security screening in airports and sensitive facilities. Due to their ability to detect metallic and nonmetallic objects underclothing, mm-wave scanners provide a noninvasive method of screening individuals without radiation risks associated with X-rays (Kemp, 2007; Wang et al., 2019).

Similarly, in surveillance, mm-wave radar systems offer enhanced imaging and motion detection capabilities, even in low-visibility conditions such as fog or smoke. These systems are also being explored for use in smart home security, autonomous drones, and perimeter monitoring in critical infrastructure.

Millimeter-wave communication is no longer confined to laboratory research or niche applications. It is becoming a critical component of modern communication infrastructure and is poised to redefine how we interact with technology across multiple domains. From ultra-fast mobile networks and smart cities to medical diagnostics and autonomous vehicles, the capabilities of mm-wave frequencies are unlocking a new realm of possibilities. While challenges related to propagation, penetration, and component design persist, ongoing research and innovation are rapidly addressing these issues. As we look toward the future, mm-wave communication stands out not only as a solution to current spectrum congestion but also as a catalyst for the next wave of technological transformation.

2.3 Impairments in mm-Wave Antennas and Communication Systems

Millimeter-wave technology is becoming increasingly important as wireless communication systems advance toward 5G and beyond. Operating typically in the 30–300 GHz range, mm-wave frequencies offer wide bandwidth and high data rates. However, these benefits come with a set of unique challenges that can significantly impair performance (Wang et al., 2018). This section outlines the most prominent impairments affecting mm-wave communication systems and antennas, and includes the mathematical tools engineers use to quantify them. Key impairments in mm-wave communication systems are shown in Figure 2.2.

2.3.1 High Free-space Path Loss

At mm-wave frequencies, signals experience much greater FSPL than at lower frequencies. This is primarily because the wavelength is smaller, leading to faster energy dispersion over distance. The FSPL can be calculated using the Friis transmission equation:

$$\text{FSPL (dB)} = 20\log_{10}(d) + 20\log_{10}(f) + 20\log_{10}\left(\frac{4\pi}{c}\right), \tag{2.2}$$

where:

- d is the distance between the transmitter and receiver (in meters),
- f is the frequency (in Hz),
- c is the speed of light ($\approx 3 \times 10^8$ m/s).

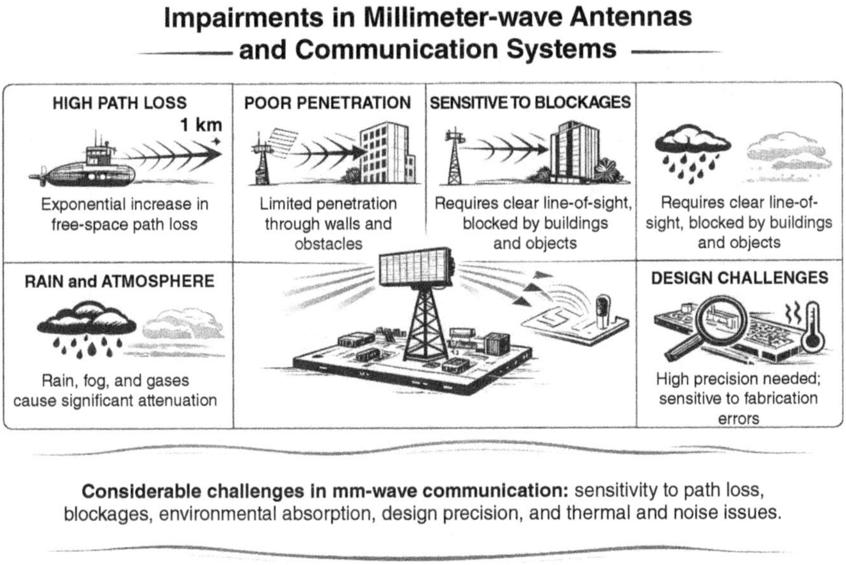

Figure 2.2 Key impairments in mm-wave communication systems.

As frequency increases, the path loss increases logarithmically. For instance, a signal at 60 GHz will experience significantly more loss than one at 2.4 GHz over the same distance.

2.3.2 Poor Penetration Through Obstacles

Unlike microwave signals, mm-waves struggle to penetrate walls, glass, and even human tissue. Most materials act as either reflectors or absorbers at these high frequencies, reducing the signal strength after passing through. This severely limits mm-wave performance in indoor or urban environments. A basic attenuation model for walls can be expressed as:

$$L = L_0 + \alpha d, \tag{2.3}$$

where:

- L is the total path loss (dB),
- L_0 is the initial penetration loss at the material boundary,
- α is the attenuation constant (dB/m),
- d is the thickness of the material (m).

The values of L_0 and α vary based on the material, but tend to be significantly higher for mm-waves.

2.3.3 Sensitivity to Line-of-sight and Blockages

Millimeter-wave signals rely heavily on a clear LOS path. When LOS is obstructed, signal strength may drop sharply. Nonline-of-sight (NLOS) communication is possible through reflection and scattering, but these secondary paths are weaker and often unstable. Reflection loss, R, at a dielectric interface is given by the Fresnel equations. For perpendicular incidence:

$$R = \left| \frac{n_1 - n_2}{n_1 + n_2} \right|^2, \tag{2.4}$$

where n_1 and n_2 are the refractive indices of the incident and transmission media, respectively.

2.3.4 Rain, Fog, and Atmospheric Absorption

Environmental conditions have a pronounced effect on mm-wave propagation. Atmospheric gases, especially oxygen and water vapor, absorb mm-wave energy, and weather conditions like rain and fog scatter the signal. The specific attenuation due to rain, A_R, can be estimated using ITU models:

$$A_R = kR^{\alpha}, \tag{2.5}$$

where:

- A_R is the rain attenuation (dB/km),
- R is the rain rate (mm/h),
- k and α are frequency- and polarization-dependent constants.

At 60 GHz, oxygen absorption can lead to atmospheric losses exceeding 10 dB/km, limiting effective range for outdoor links.

2.3.5 Challenges in Antenna and Circuit Design

At higher frequencies, even minor fabrication inaccuracies can cause significant performance degradation. Tolerances in substrate thickness, conductor width, and dielectric constant have to be extremely tight. Losses from metal conductivity and dielectric materials become critical. The conductor loss, P_c, for a transmission line can be estimated as:

$$P_c = \frac{1}{2} R_s \int |H_t|^2 \, dl, \tag{2.6}$$

where:

- R_s is the surface resistance,
- H_t is the tangential magnetic field,
- dl is the differential length element along the conductor surface.

Surface resistance increases with frequency due to the skin effect:

$$R_s = \sqrt{\frac{\pi f \mu}{\sigma}}, \tag{2.7}$$

where:

- f is the frequency,
- μ is the permeability of the conductor,
- σ is the conductivity.

2.3.6 Beamforming and Mobility Limitations

To mitigate high path loss, mm-wave systems often use beamforming antenna systems. While effective, these systems are complex and can struggle with real-time tracking of mobile users. Maintaining alignment between beams is a challenge in dynamic environments, especially if obstructions are present or users move rapidly. Fast beam switching requires sophisticated algorithms and often increases power consumption and hardware complexity.

2.3.7 Thermal and Noise Issues

As mm-wave devices operate at high frequencies, they generate more heat in smaller packages. This thermal concentration can affect device longevity and performance. Moreover, high-frequency components are more susceptible to phase noise and nonlinear distortion. The overall SNR becomes a limiting factor for performance. The Shannon–Hartley theorem provides an upper bound on channel capacity:

$$C = B \log_2\left(1 + \frac{S}{N}\right), \tag{2.8}$$

where:

- C is the channel capacity (bits/s),
- B is the bandwidth (Hz),
- S is the signal power,
- N is the noise power.

At mm-wave bands, both S and N are affected by system design, making SNR optimization a central design challenge.

Despite their potential for enabling high-speed, high-capacity communication, mm-wave systems are inherently prone to a wide range of impairments. These include high path loss, sensitivity to obstacles, environmental absorption, and complex hardware requirements. Successfully implementing mm-wave communication requires careful system design, advanced fabrication techniques, and continuous adaptation to real-world conditions. By understanding and addressing these impairments, researchers and engineers can unlock the full potential of this exciting frequency range.

2.4 Design Strategies for mm-Wave Antennas

Designing antennas for mm-wave applications requires a shift in approach compared to conventional microwave antenna design. As the frequency increases, so do the constraints and subtleties in every part of the design process—from material selection to layout optimization and packaging. While mm-wave systems offer impressive bandwidth and enable miniaturized components, these benefits come with challenges that demand thoughtful strategies to address. A summary of key design strategies for mm-wave antennas is shown in Figure 2.3.

2.4.1 Material Considerations

At mm-wave frequencies, even small material imperfections can affect performance. The dielectric substrate must have a low loss tangent and consistent dielectric constant across the operating frequency range. Materials like Rogers RT/duroid, liquid crystal polymer (LCP), and silicon-based substrates are commonly used. A consistent dielectric profile helps reduce signal distortion and phase instability, both of which become prominent at higher frequencies.

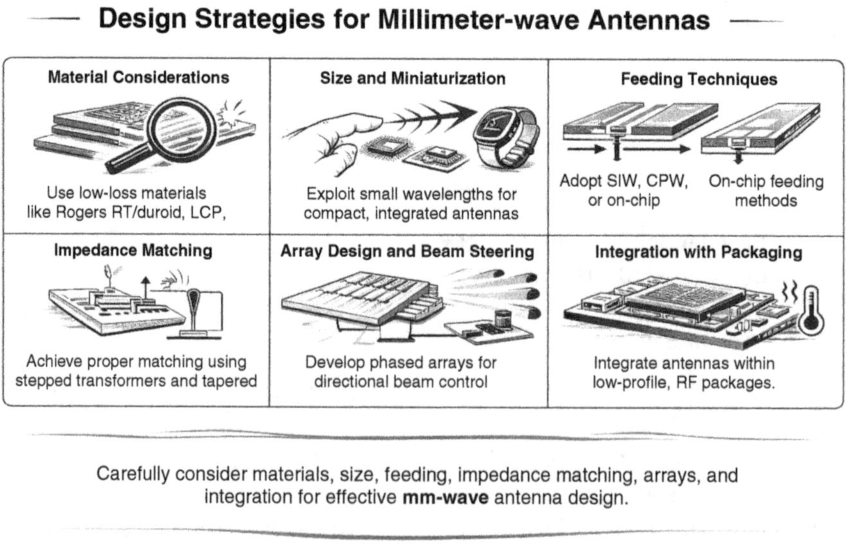

Figure 2.3 Summary of key design strategies for mm-wave antennas.

2.4.2 Size and Miniaturization

One advantage of mm-wave design is the naturally small wavelength, which allows antennas to be significantly smaller than their microwave counterparts. This is useful for applications where space is at a premium, such as in wearables, chipsets, or biomedical implants. However, smaller antennas often suffer from narrower bandwidth and reduced efficiency. Designers often compensate by integrating arrays or employing bandwidth-enhancing techniques such as slot loading, dielectric resonators (DRAs), or multi-resonant structures.

2.4.3 Feeding Techniques

Traditional feeding methods like coaxial probes or microstrip lines can introduce significant parasitics or become impractical at mm-wave scales. Coplanar waveguides (CPW), substrate-integrated waveguides (SIW), and on-chip feeding mechanisms are more suitable due to their low loss and compact form. SIW structures, in particular, offer good performance while maintaining planar integration, which is ideal for monolithic designs.

2.4.4 Impedance Matching and Bandwidth Enhancement

Impedance matching becomes trickier at mm-wave frequencies due to narrow fabrication tolerances and variations in connector quality. Small mismatches can lead to significant reflection losses. To tackle this, stepped-impedance transformers, tapered feed lines, and wideband matching networks are often integrated into the design. Bandwidth can also be enhanced by using parasitic elements or by introducing stubs and notches that support multiple resonant modes.

2.4.5 Array Design and Beam Steering

To compensate for higher path loss and achieve directional communication, antenna arrays are widely used at mm-wave bands. For instance, phased arrays, in particular, allow for beam steering, which helps maintain strong communication links even when the transmitter or receiver is moving. Design strategies here focus on minimizing mutual coupling, optimizing inter-element spacing (typically around half a wavelength), and ensuring good amplitude and phase balance across elements.

2.4.6 Integration with Packaging

Antenna performance at mm-wave frequencies is highly sensitive to the surrounding environment. This makes packaging a critical part of the design, not an afterthought. Antennas are often integrated into the same substrate as other RF components to reduce losses and minimize parasitic effects. Low-profile, conformal packaging is preferred, and designers must account for enclosures, metallic frames, and even proximity to the human body in some applications, all of which can detune or degrade the antenna's performance.

2.4.7 Simulation and Fabrication Constraints

Modern design strategies heavily rely on accurate full-wave simulation tools like HFSS, CST, or ADS. At mm-wave frequencies, simulation accuracy is critical due to the wavelength-scale dimensions involved. Yet, designs must also be robust against manufacturing tolerances, especially in mass-produced or on-chip systems. To address this, designers often run parametric sweeps and tolerance analysis as part of the design cycle to ensure consistent real-world performance.

In short, designing for mm-wave frequencies is not just a matter of scaling down existing microwave techniques. It demands a careful balance between performance, manufacturability, and system integration. From choosing the right materials to navigating tight tolerances and integrating with circuits, every decision counts. With the right design strategies, mm-wave antennas can unlock unprecedented possibilities in high-speed wireless communication, sensing, and emerging applications.

2.5 Fabrication and Measurement Techniques for mm-Wave Antennas

Designing a mm-wave antenna is only half the battle. Once the simulations are complete and performance goals are met in software, the real-world realization begins. Fabrication and measurement are crucial steps that bring theory to practice, and both come with their own unique set of challenges, especially at such high frequencies where even tiny imperfections can have a significant impact on performance.

2.5.1 Fabrication Techniques

At mm-wave frequencies, the physical size of the antenna becomes extremely small, often on the order of a few millimeters or less. This miniaturization demands a higher level of precision in fabrication compared to microwave or lower frequency antennas. The methods used must ensure both structural integrity and electrical accuracy. An overview of key fabrication techniques for mm-wave antennas is shown in Figure 2.4.

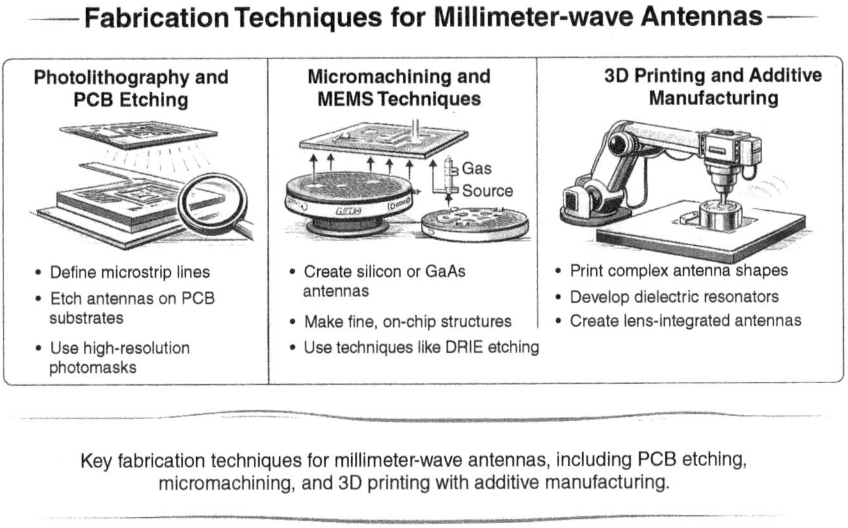

Figure 2.4 Overview of key fabrication techniques for mm-wave antennas.

2.5.1.1 Photolithography and PCB Etching

One of the most commonly used methods for fabricating planar mm-wave antennas is standard printed circuit board (PCB) etching, particularly when using substrates like Rogers RO4003C or RT/duroid 5880. This technique, adapted from the semiconductor industry, is employed to define intricate patterns with tight tolerances.

The process typically involves:

- Coating the substrate with a photosensitive resist.
- Exposing it to UV light through a patterned mask.
- Developing the exposed areas to reveal the circuit layout.
- Etching away the unwanted metal.

High-resolution photomasks are essential, especially when dealing with microstrip lines and slots in the sub-millimeter range. Variations in trace width as small as 10 μm can lead to mismatched impedances, resulting in undesirable reflections and losses.

2.5.1.2 Micromachining and MEMS Techniques

For even finer structures, particularly in on-chip or integrated systems, micromachining is used. Techniques such as deep reactive ion etching (DRIE) or bulk silicon etching are adopted from MEMS fabrication processes. These enable high precision and are ideal for antennas integrated into silicon or GaAs wafers.

Micromachining is beneficial in applications like:

- SIW antennas.
- High-frequency filter-antenna combinations.
- On-chip phased arrays and reconfigurable structures.

While these processes are more expensive and require cleanroom facilities, the dimensional accuracy and repeatability they offer are often necessary at mm-wave frequencies.

2.5.1.3 3D Printing and Additive Manufacturing

Recently, additive manufacturing has made its way into mm-wave antenna prototyping. 3D printing with high-resolution polymer or metallic materials allows for the creation of nonplanar and conformal structures. While surface roughness and conductivity remain concerns, post-processing steps such as metallization or polishing can help mitigate these issues.

3D-printed antennas are particularly useful for:

- Dielectric resonator antennas.
- Lens-integrated antenna structures.
- Customized enclosures with embedded antennas.

2.5.2 Measurement Techniques

Once the antenna is fabricated, the next step is to evaluate its performance. Unlike simulation tools, which offer idealized predictions, measurements reflect the realities of manufacturing tolerances, material inconsistencies, and connector losses. At mm-wave frequencies, precision in measurement is just as vital as in fabrication. Measurement techniques for mm-wave antennas are illustrated in Figure 2.5.

2.5.2.1 S-parameter Measurements Using Vector Network Analyzers

The first metric most engineers check is the return loss (S_{11}) and transmission coefficient (S_{21}, if applicable). These are measured using a vector network analyzer (VNA) capable of operating at mm-wave frequencies, typically up to 110 GHz or more.

VNAs use calibrated ports to send and receive signals. To achieve accurate results, the test setup must be carefully calibrated using standards like short-open-load-through (SOLT) or thru-reflect-line (TRL). Calibration ensures that losses from cables, connectors, and adapters are de-embedded from the measurements.

$$S_{11}(\text{dB}) = 20 \log\left|\frac{V_{\text{reflected}}}{V_{\text{incident}}}\right|.$$

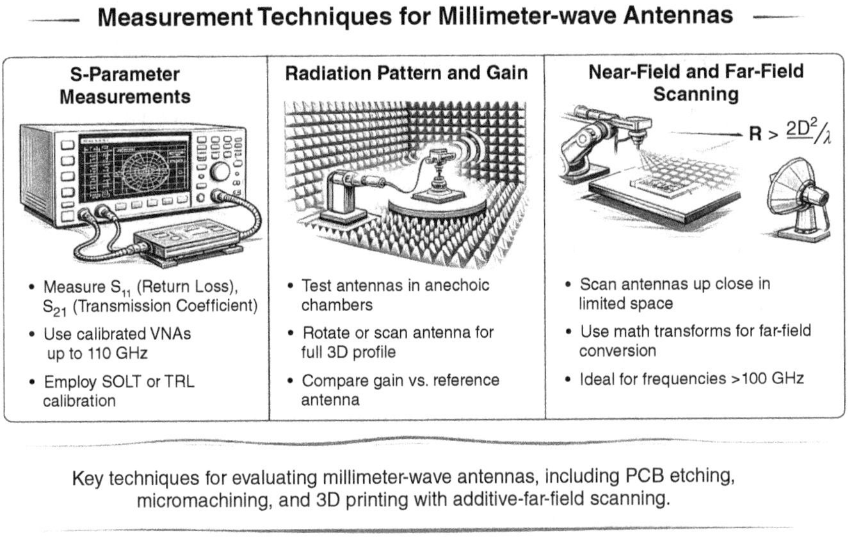

Figure 2.5 Measurement techniques for mm-wave antennas.

Any mismatch, including minor variations in SMA or waveguide connectors, can introduce ripples and false readings. Therefore, precise calibration and high-quality connectors are critical.

2.5.2.2 Radiation Pattern and Gain Measurements

Radiation pattern measurements involve placing the antenna under test (AUT) in an anechoic chamber, isolated from reflections and external noise. A reference antenna is used to measure the radiation in both the azimuth and elevation planes.

The antenna is rotated (or the probe is moved) through 360° to obtain a full 3D profile. This reveals the main lobe, sidelobes, back lobe, and beamwidth.

The antenna gain can be calculated using the comparison method:

$$G_{\mathrm{AUT}} = G_{\mathrm{ref}} + 20\log\left(\frac{V_{\mathrm{AUT}}}{V_{\mathrm{ref}}}\right),$$

where G_{ref} is the known gain of the reference antenna, and V_{AUT} and V_{ref} are the received voltages of the AUT and reference antenna, respectively.

2.5.2.3 Near-field and Far-field Scanning

Depending on the chamber size and frequency range, the near-field scanning may be used to measure mm-wave antennas, especially when the far-field distance (calculated by $R > 2D^2/\lambda$) becomes unmanageable. Near-field measurements can be transformed mathematically into far-field patterns using Fourier transform techniques.

2.5.3 Challenges and Best Practices

Some key challenges in mm-wave measurements include:

- **Connector repeatability:** Even slight variations in torque can affect results.
- **Cable bending:** Changing the cable shape between calibration and measurement can introduce phase errors.
- **Temperature stability:** mm-wave performance can drift due to thermal expansion or contraction.
- **Alignment:** Misalignment between waveguide flanges can result in large insertion losses.

To overcome these, it is important to:

- Use torque wrenches and high-quality adapters.
- Minimize cable movements after calibration.
- Keep the environment stable and avoid airflow or temperature gradients.
- Conduct multiple measurements to ensure repeatability.

Fabrication and measurement of mm-wave antennas are intricate processes that require a careful balance of precision engineering and practical understanding. As we push toward higher frequencies with 5G, 6G, and beyond, mastering these techniques becomes not just beneficial but essential. Engineers must be meticulous in fabrication, detail-oriented in calibration, and adaptive in handling real-world deviations. Only then can the true potential of mm-wave antennas be fully realized in practical communication systems.

2.6 Conclusion

In this chapter, we have introduced the fundamental aspects of mm-wave wireless communication systems. Beginning with the structure of the RF spectrum, we have explored how different frequency ranges are categorized and how mm-wave frequencies, typically ranging from 30 to 300 GHz, stand out due to their potential to support extremely high data rates and compact antenna design.

We have then discussed various practical applications where mm-wave technology is proving to be transformative. From ultra-fast 5G connectivity and autonomous vehicle radar systems to high-resolution imaging and next-generation wearable health devices, these use cases demonstrate the growing importance of mm-wave communication in today's connected world.

Despite their promise, mm-wave systems face significant challenges. Issues such as high path loss, atmospheric absorption, limited diffraction around obstacles, and increased system noise complicate deployment and performance. These impairments highlight the need for robust system design and careful component integration.

To mitigate such challenges, we have presented design strategies that consider not only the electrical performance but also fabrication feasibility, cost, and mechanical robustness. Simulations using full-wave solvers, impedance matching techniques, material selection, and feeding network design are all critical considerations during the development process.

Fabrication and measurement techniques also play a central role in bridging the gap between theory and real-world performance. With shrinking wavelengths, the precision in manufacturing becomes increasingly critical. Techniques like PCB milling, photolithography, and 3D printing, along with accurate measurement setups using VNAs and anechoic chambers, ensure that the final antenna performs as expected.

In summary, this chapter has provided a holistic overview of mm-wave antenna systems, covering everything from spectrum fundamentals to hands-on implementation challenges. These insights build the necessary groundwork for the more specialized antenna structures and configurations discussed in the subsequent chapters.

References

Michael Cheffena. Industrial wireless communications over the millimeter wave spectrum: opportunities and challenges. *IEEE Communications Magazine*, 54(9):66–72, 2016. https://doi.org/10.1109/MCOM.2016.7565190.

Zhi Ning Chen, Duixian Liu, Hisamatsu Nakano, Xianming Qing, and Thomas Zwick. *Handbook of Antenna Technologies*. Springer Publishing Company, Incorporated, 2016.

Gayatri Chittimoju and Usha Devi Yalavarthi. A comprehensive review on millimeter waves applications and antennas. In *Journal of Physics: Conference Series*, vol. 1804. IOP Publishing, 2021.

Junil Choi, Vutha Va, Nuria Gonzalez-Prelcic, Robert Daniels, Chandra R. Bhat, and Robert W. Heath. Millimeter-wave vehicular communication to support massive automotive sensing. *IEEE Communications Magazine*, 54(12):160–167, 2016. https://doi.org/10.1109/MCOM.2016.1600071CM.

Chinh H Doan, Sohrab Emami, David A Sobel, Ali M Niknejad, and Robert W Brodersen. Design considerations for 60 GHz CMOS radios. *IEEE Communications Magazine*, 42(12):132–140, 2004.

Sourjya Dutta, Marco Mezzavilla, Russell Ford, Menglei Zhang, Sundeep Rangan, and Michele Zorzi. Mac layer frame design for millimeter wave cellular system. In *2016 European Conference on Networks and Communications (EuCNC)*, pages 117–121. IEEE, 2016.

Michael C. Kemp. Millimetre wave and terahertz technology for detection of concealed threats—a review. In *2007 Joint 32nd International Conference on*

Infrared and Millimeter Waves and the 15th International Conference on Terahertz Electronics, pages 647–648, 2007. https://doi.org/10.1109/ICIMW.2007.4516664.

Aida Vera Lopez, Andrey Chervyakov, Greg Chance, Sumit Verma, and Yang Tang. Opportunities and challenges of mmwave NR. *IEEE Wireless Communications*, 26(2):4–6, 2019. https://doi.org/10.1109/MWC.2019.8700132.

T. Manabe, K. Sato, H. Masuzawa, K. Taira, T. Ihara, Y. Kasashima, and K. Yamaki. Polarization dependence of multipath propagation and high-speed transmission characteristics of indoor millimeter-wave channel at 60 GHz. *IEEE Transactions on Vehicular Technology*, 44(2):268–274, 1995. https://doi.org/10.1109/25.385918.

Shahid Mumtaz, Jonathan Rodriguez, and Linglong Dai. *MmWave Massive MIMO: A Paradigm for 5G*. Academic Press, 2016.

Aqeel Hussain Naqvi and Sungjoon Lim. Review of recent phased arrays for millimeter-wave wireless communication. *Sensors*, 18(10):3194, 2018.

Zhao Qingling and Jin Li. Rain attenuation in millimeter wave ranges. In *2006 7th International Symposium on Antennas, Propagation & EM Theory*, pages 1–4. IEEE, 2006.

Theodore S Rappaport, Yunchou Xing, George R MacCartney, Andreas F Molisch, Evangelos Mellios, and Jianhua Zhang. Overview of millimeter wave communications for fifth-generation (5G) wireless networks—with a focus on propagation models. *IEEE Transactions on Antennas and Propagation*, 65(12):6213–6230, 2017.

Ad CF Reniers, Ulf Johannsen, and AB Smolders. Antenna-in-package measurements. *Antenna-in-Package Technology and Applications*, pages 115–145, 2020.

David M. Sheen, Douglas L. McMakin, and Thomas E. Hall. Near field imaging at microwave and millimeter wave frequencies. In *2007 IEEE/MTT-S International Microwave Symposium*, pages 1693–1696, 2007. https://doi.org/10.1109/MWSYM.2007.380033.

Alexey V. Shvetsov and Saeed Hamood Alsamhi. When holographic communication meets metaverse: Applications, challenges, and future trends. *IEEE Access*, 12:197488–197515, 2024. https://doi.org/10.1109/ACCESS.2024.3514576.

Jakob Struye, Sam Van Damme, Nabeel Nisar Bhat, Arno Troch, Barend Van Liempd, Hany Assasa, Filip Lemic, Jeroen Famaey, and Maria Torres Vega. Toward interactive multi-user extended reality using millimeter-wave networking. *IEEE Communications Magazine*, 62(8):54–60, 2024. https://doi.org/10.1109/MCOM.001.2300804.

Berke Tezergil and Ertan Onur. Wireless backhaul in 5G and beyond: Issues, challenges and opportunities. *IEEE Communications Surveys & Tutorials*, 24(4):2579–2632, 2022. https://doi.org/10.1109/COMST.2022.3203578.

Fritzi Topfer and Joachim Oberhammer. Millimeter-wave tissue diagnosis: The most promising fields for medical applications. *IEEE Microwave Magazine*, 16(4):97–113, 2015. https://doi.org/10.1109/MMM.2015.2394020.

Peng Wang, Yonghui Li, Lingyang Song, and Branka Vucetic. Multi-gigabit millimeter wave wireless communications for 5G: From fixed access to cellular networks. *IEEE Communications Magazine*, 53(1):168–178, 2015.

Xiong Wang, Linghe Kong, Fanxin Kong, Fudong Qiu, Mingyu Xia, Shlomi Arnon, and Guihai Chen. Millimeter wave communication: A comprehensive survey. *IEEE Communications Surveys & Tutorials*, 20(3):1616–1653, 2018. https://doi.org/10.1109/COMST.2018.2844322.

Zhongmin Wang, Tianying Chang, and Hong-Liang Cui. Review of active millimeter wave imaging techniques for personnel security screening. *IEEE Access*, 7:148336–148350, 2019. https://doi.org/10.1109/ACCESS.2019.2946736.

Muhammad Zada, Izaz Ali Shah, and Hyoungsuk Yoo. Integration of sub-6-GHz and mm-wave bands with a large frequency ratio for future 5G MIMO applications. *IEEE Access*, 9:11241–11251, 2021. https://doi.org/10.1109/ACCESS.2021.3051066.

3

Millimeter-wave Planar Antennas

3.1 Introduction

As wireless communication technologies continue to evolve, there is a growing need for faster data rates, lower latency, and better spectrum utilization. One major step toward achieving these goals is the transition to millimeter wave (mm-wave) frequencies, typically defined as ranging from 30 to 300 GHz. These higher frequencies offer wide bandwidths that can support high-capacity communication systems, making them ideal for fifth-generation (5G), sixth-generation (6G), automotive radar, satellite links, and other advanced wireless technologies (Aparna et al., 2024; Tiwari et al., 2024).

At such frequencies, the design and integration of antennas become increasingly important. The antenna serves as a critical interface between the electronic circuitry and the electromagnetic environment. In the case of mm-wave systems, the short wavelength allows antennas to be much smaller, which opens up opportunities for more compact designs and densely packed antenna arrays. However, this also brings new challenges that must be carefully addressed in both design and implementation (Li et al., 2024).

Because of their miniaturized geometry, non-bulky architecture, and easy manufacturing process, they are highly regarded for mm-wave wireless systems. Due to their two-dimensional (2D) topology, they can be directly printed on the printed circuit board (PCB) using laser etching or chemical etching. The production of such antennas is easy. Moreover, they are straightforward to integrate with other radio frequency (RF) components. In addition, these antennas can be fabricated on the PCB with other RF components in an integrated form (Raj and Mandal, 2024). An example of a coplanar waveguides (CPW)-fed microstrip patch antenna is shown in Figure 3.1.

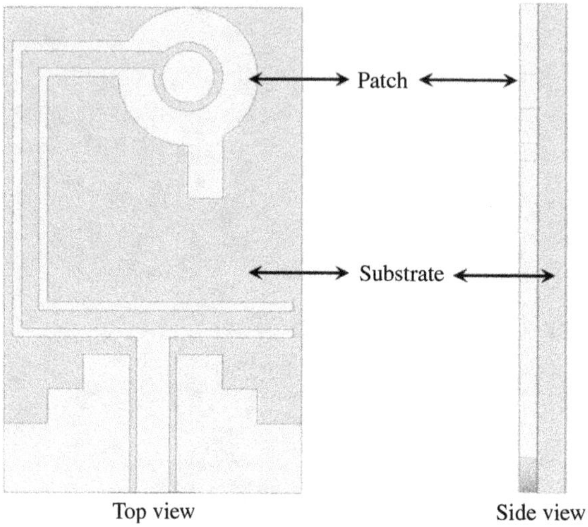

Figure 3.1 CPW-fed microstrip planar antenna.

3.1.1 Why Planar Antennas for mm-Wave?

The wavelengths of mm-wave antennas are shorter, which allows the radiating antenna to be very small in size. Due to this advantage, designing a high-gain antenna array, working at mm-wave frequency range, becomes relatively easy for very compact devices. The antenna array architecture is specifically important in designing high-gain beamforming and MIMO systems. Such systems can deliver data with high channel capacity over a wide range.

Unlike 3D antennas, these antennas can be made using standard fabrication methods like photolithography (chemical etching) or laser etching. This makes them ideal for large-scale production, offering a scalable and cost-effective solution for the high-frequency demands of modern technology.

3.1.2 Design Considerations and Challenges

Designing planar antennas for mm-wave applications is not simply a matter of shrinking existing designs. As frequency increases, the electrical behavior of materials changes, and new challenges emerge. For example, dielectric losses become more significant, and conductor losses increase due to the reduced skin depth. Even small variations in fabrication tolerances can greatly affect the performance.

The feeding methods for mm-wave antennas must also be adapted for operation at mm-wave. While edge or probe feeding methods are common at lower frequencies, they often result in increased parasitic effects at mm-wave bands. Instead, techniques such as CPWs, proximity coupling, and substrate integrated waveguides (SIWs) are preferred to maintain efficient and broadband performance. An example of a CPW-fed microstrip patch antenna is shown in Figure 3.2. An example of an SIW-fed antenna is shown in Figure 3.3.

The choice of substrate material becomes more critical as well. Low-loss, high-frequency substrates are needed to minimize signal degradation and maintain performance. Parameters such as dielectric constant, loss tangent, and thermal stability must be considered carefully in the design phase.

3.1.3 Applications of Planar mm-Wave Antennas

Planar antennas have been widely used across a variety of mm-wave wireless applications. For instance, in 5G and emerging 6G networks, these antennas enable high-data-rate links in user equipment and small cells. In automotive systems, they support radar-based functions such as collision avoidance and adaptive cruise

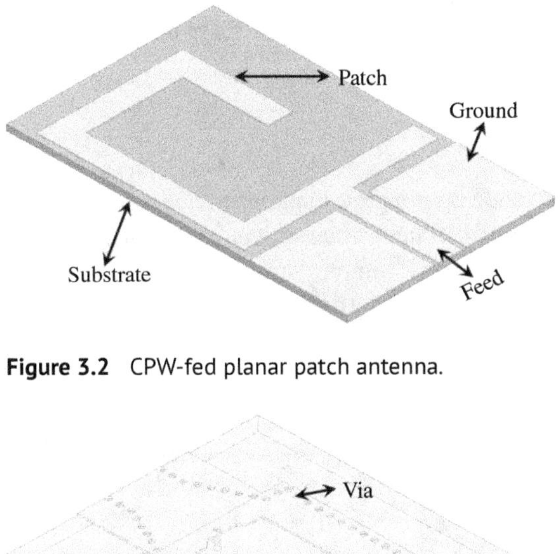

Figure 3.2 CPW-fed planar patch antenna.

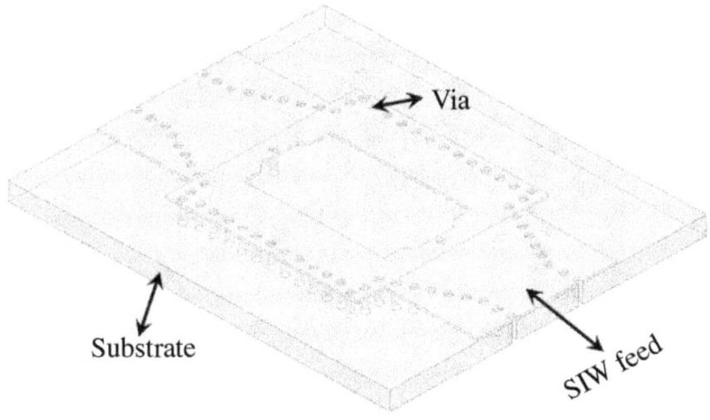

Figure 3.3 SIW-fed 5G antenna.

control. In satellite communications, especially for low Earth orbit (LEO) systems, they are used in electronically steerable arrays for dynamic beam tracking.

Unlike 3D antennas, these antennas can be made using standard fabrication methods like photolithography (chemical etching) or laser etching. This makes them ideal for large-scale production, offering a scalable and cost-effective solution for the high-frequency demands of modern technology.

In addition, these antennas can be used in other applications such as wearable sensors and electronics, wearable and free space monitoring devices, and cell/tissue imaging. All these devices require a very compact antenna geometry in planar form, which is possible with mm-wave technology.

3.2 Microstrip Antennas

Because of their extremely low size, simplicity in integration with other RF circuits, and compatibility with PCB technology, microstrip antennas are among the most commonly used antennas. Their usefulness extends well into the mm-wave frequency range, and these characteristics make them perfect for small wireless devices, especially at mm-wave bands.

3.2.1 Basic Structure and Operation

Typically, a microstrip antenna is made up of a ground plane on one side and a thin metallic patch on top of a dielectric substrate. Depending on the application, the patch, which serves as the radiating element, can have a variety of shapes, including circular, rectangular, or more intricate geometries. For instance, a rectangular microstrip patch antenna is shown in Figure 3.4.

When fed with an RF signal, the patch resonates and radiates electromagnetic energy. The dimensions of the patch (length and width in case of rectangular or

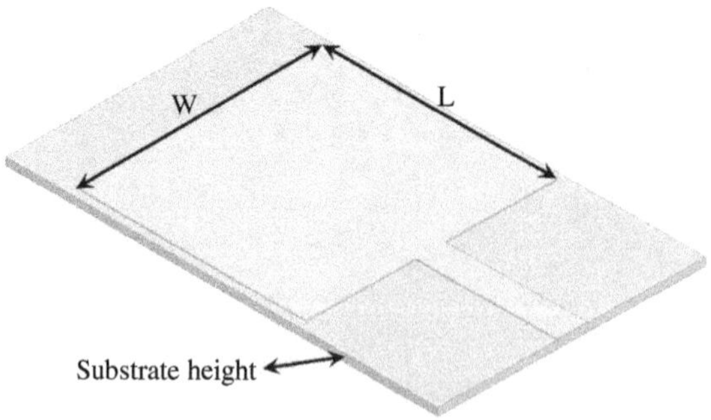

Figure 3.4 Rectangular microstrip patch antenna.

square patch), height of the substrate, and permittivity of the substrate determine the resonance of the antenna. According to the shape of the patch, a different formula can be used. For example, for a rectangular patch antenna, the following equations can be used (Balanis, 2016):

Width of the patch (W)

$$W = \frac{c}{2f_r}\sqrt{\frac{2}{\varepsilon_r + 1}}, \tag{3.1}$$

where

- c is the speed of light (3×10^8 m/s)
- f_r is the resonant frequency in Hz
- ε_r is the relative permittivity of the substrate

Effective dielectric constant (ε_{eff})

$$\varepsilon_{\text{eff}} = \frac{\varepsilon_r + 1}{2} + \frac{\varepsilon_r - 1}{2}\left(1 + 12\frac{h}{W}\right)^{-1/2}, \tag{3.2}$$

where h is the substrate height.

Effective length (L_{eff})

$$L_{\text{eff}} = \frac{c}{2f_r\sqrt{\varepsilon_{\text{eff}}}}, \tag{3.3}$$

Length extension due to fringing fields (ΔL)

$$\Delta L = 0.412h \cdot \frac{(\varepsilon_{\text{eff}} + 0.3)\left(\frac{W}{h} + 0.264\right)}{(\varepsilon_{\text{eff}} - 0.258)\left(\frac{W}{h} + 0.8\right)}, \tag{3.4}$$

Actual length of the patch (L)

$$L = L_{\text{eff}} - 2\Delta L. \tag{3.5}$$

3.2.2 Advantages at mm-Wave Frequencies

Because of the shorter wavelength at mm-wave frequencies, the patch's actual size is typically only a few millimeters. This small size is useful for combining several antennas into arrays, which are necessary for methods like spatial diversity and beamforming.

Microstrip antennas can be made using conventional photolithographic techniques and are planar in addition to being tiny. Because of this, they can be produced in large quantities and are very appealing for commercial mm-wave goods such as wearable technology, vehicle radar, and 5G user interfaces.

3.2.3 Design Considerations and Challenges

Designing microstrip antennas at mm-wave frequencies introduces several unique challenges. The main design considerations and challenges for mm-wave microstrip antennas are illustrated in Figure 3.5.

- **Material Losses:** As frequency increases, both dielectric and conductor losses become more pronounced. Low-loss substrates and high-conductivity metals must be used to maintain efficiency.
- **Narrow Bandwidth:** Traditional patch antennas offer narrow bandwidths, which can be insufficient for high-data-rate mm-wave applications. Techniques such as stacking patches or using parasitic elements are employed to widen the bandwidth.
- **Feeding Techniques:** Increased parasitic effects make edge feeding and probe feeding less effective. Other methods, such as proximity coupling and inset feeds, are frequently chosen.
- **Fabrication Precision:** At mm-wave frequencies, very tight tolerances are needed. Performance deterioration and detuning may arise from small variations.

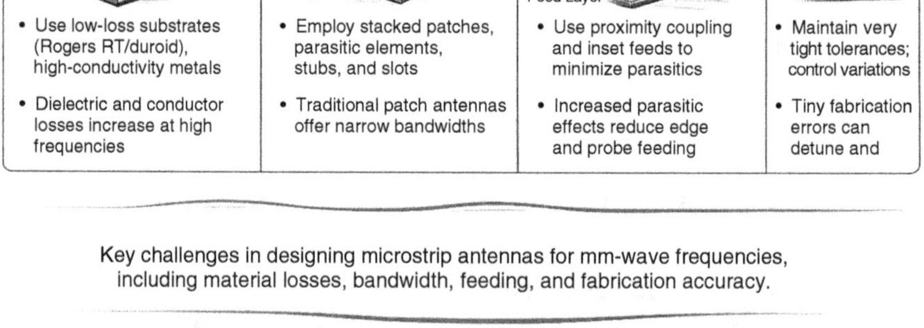

Figure 3.5 Main design considerations and challenges for millimeter-wave microstrip antennas.

3.2.4 Applications

In mm-wave communication systems, such as satellite terminals, 5G/6G mobile devices, V2X, V2V, and Internet of Things sensors, microstrip antennas have been widely utilized. They are an essential component of contemporary wireless technologies due to their portability and adaptability.

3.2.5 Design Example

The authors of Hussain and Kim (2022) have proposed a patch antenna that operates at both microwave and mm-wave frequency bands. It covers the 2.5, 3.5, 5.5, and 7.5 GHz bands (microwave bands) and also operates at mm-wave frequencies ranging from 23 to 31 GHz. The antenna demonstrates superior performance in terms of radiation characteristics. Specifically, it exhibits a high radiation efficiency of more than 95% and a stable gain of over 2.5 dBi in the microwave bands and 6.5 dBi in the mm-wave band. The geometry of the proposed antenna is shown in Figure 3.6. The antenna has a very compact size of $16 \times 21\ \text{mm}^2$.

Figure 3.7 shows the progressive evolution of integrated microwave and mm-wave antenna design. Initially, a triangular patch is designed, where the width is calculated using the following equation.

$$w_p \approx \frac{c}{4f_{\text{res}}\sqrt{\frac{\varepsilon_r+1}{2}}}. \tag{3.6}$$

Once the radiator width is calculated, a bent stub is added to elongate the current path. The design is gradually advanced towards a multiband integrated configuration working at both microwave and mm-wave frequencies. In the first step (Antenna-I), only a triangular patch along with a bent stub is simulated. In this iteration, the antenna has resonances of 2.5 and 8.5 GHz. In this design, the patch generates 8.5 GHz resonance and bent stub generates 2.5 GHz resonance. The resonance of Antenna-I has a limited bandwidth. This basic configuration serves as a foundational step but is clearly insufficient for modern communication systems that require broader and more versatile frequency support.

As the design moves to Antenna-2, another $\lambda/4$ stub is added to the existing design. The length of the stubs is calculated so it resonates at 3.5 GHz. The following formula has been used to calculate the length of the additional stub.

$$L_{3.5\text{GHz}} = l_2 + w_2 = \frac{\lambda_{3.5\text{GHz}}}{4}. \tag{3.7}$$

By adding a second bent stub, an additional resonance is noted at 3.5 GHz, as shown in Figure 3.8. As a result, the antenna in the second iteration has resonances at 2.5, 3.5, and 8.5 GHz. The improvements lead to multiple matching dips within the sub-6 GHz region, suggesting early signs of multiband capability. However,

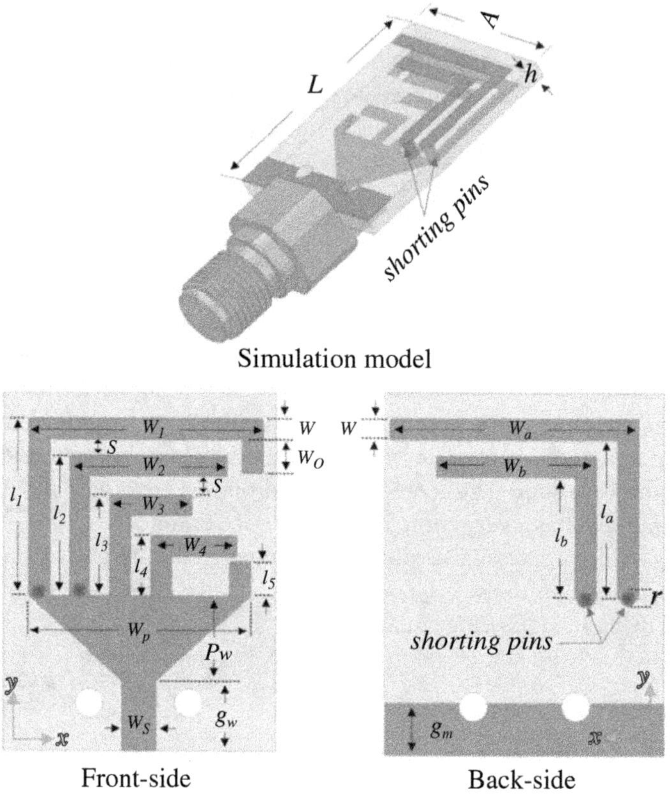

Figure 3.6 Geometry of the integrated microwave and mm-wave antenna ($A = 16$, $L = 21$, $h = 1.6$, $g_m = 3$, $w_s = 2$, $g_w = 4.1$, $w_p = 13$, $p_w = 5$, $l_1 = 10.5$, $w_1 = 13.7$, $l_2 = 8.2$, $w_2 = 9.2$, $l_3 = 5.9$, $w_3 = 4.8$, $l_4 = 3.5$, $w_4 = 5$, $l_5 = 2$, $s = 1$, $w = 1$, $w_o = 3.2$, $w_a = 14.2$, $w_b = 9.2$, $l_a = 10.5$, and $l_b = 8.2$ [unit = mm]). *Source: Hussain and Kim (2022)/IEEE/CC BY 4.0.*

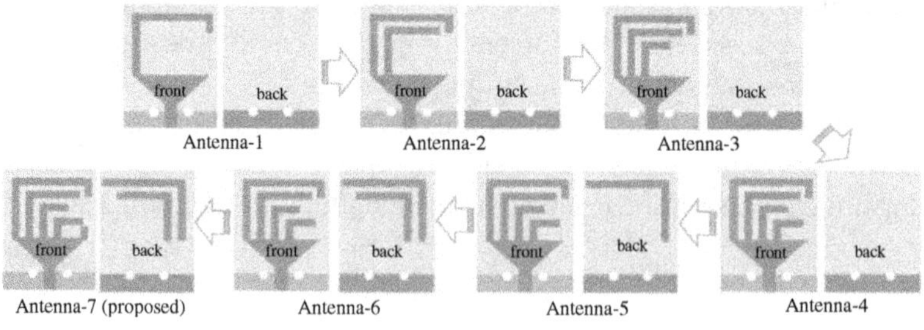

Figure 3.7 Design evolution. *Source: Hussain and Kim (2022)/IEEE/CC BY 4.0.*

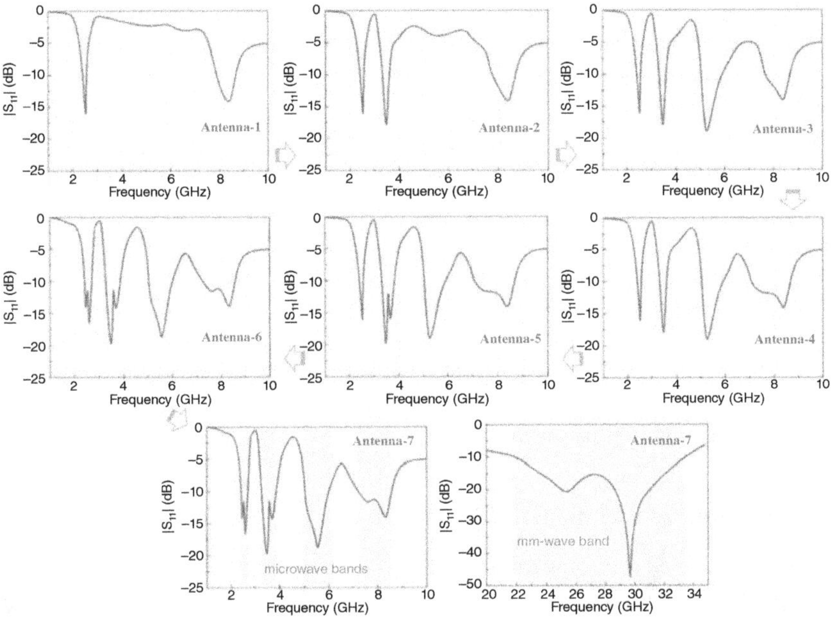

Figure 3.8 S_{11} of the design evolution. *Source: Hussain and Kim (2022)/IEEE/CC BY 4.0.*

the bandwidth at each resonance remains relatively narrow, and further refinement is necessary. In the next step (Antenna-4), another stub is added to resonate the antenna at 5.5 GHz. The stub length is calculated using the following formula.

$$L_{5.5\text{GHz}} = l_3 + w_3 = \frac{\lambda_{3.5\text{GHz}}}{4}. \tag{3.8}$$

The reflection coefficient plots show multiple deep notches, indicating stronger and more consistent impedance matching across wider frequency bands. In the next stage, two more stubs on the back are added to introduce new resonances at the lower frequency bands to improve the bandwidth.

The final version, Antenna-7, represents a major leap forward in terms of frequency coverage and matching performance. The Antenna-7 shows excellent impedance matching across multiple microwave bands, spanning around 2.5, 3.5, 5.5, 7.5, 8.5 GHz, and wideband spanning from 23 to 31 GHz. This broad and deep matching signifies a mature and well-optimized design. What makes Antenna-7 particularly remarkable is its ability to also operate efficiently in the mm-wave band. A clear and sharp resonance around 28 GHz is visible, with a very low reflection coefficient, indicating minimal power loss and strong potential for high-speed mm-wave applications.

Overall, the design evolution reflects a carefully structured and intelligent development process. Each iteration adds complexity and improves performance, transforming the antenna from a simple narrowband element into a versatile, multiband system capable of covering both microwave and mm-wave regions. This evolution is a textbook example of how iterative design and performance optimization can lead to advanced antenna solutions suitable for next-generation wireless technologies.

3.3 Aperture-coupled Antennas

Aperture-coupled antennas offer an effective alternative to direct-fed microstrip antennas, particularly in high-frequency designs where isolation, bandwidth, and integration flexibility are important.

3.3.1 Structure and Principle of Operation

An aperture-coupled antenna consists of two dielectric substrates separated by a ground plane that contains a narrow slot or aperture. The lower substrate carries the feed line, typically a microstrip or CPW, while the upper substrate supports the radiating patch. The RF energy couples from the feed line to the patch via the aperture.

This indirect feeding method helps isolate the feed circuitry from the radiating element, reducing spurious radiation and enabling more flexible design of the feed network.

3.3.2 Advantages for mm-Wave Design

Aperture-coupled designs are particularly beneficial at mm-wave frequencies due to several key features:

- **Enhanced Bandwidth:** The coupling through the aperture can be adjusted to fine-tune the antenna's bandwidth.
- **Improved Feed Isolation:** Since the feed is on a separate layer, unwanted interaction between the feed line and the radiating patch is minimized.
- **Multilayer Integration:** Allows the designer to use different materials for the radiating and feeding layers, optimizing performance and cost.
- **Compact Feed Network Design:** The feed circuitry can include active components such as amplifiers or switches without affecting the radiation characteristics.

3.3.3 Design Challenges

While offering several benefits, aperture-coupled antennas also come with their own set of challenges:

- **Alignment Precision:** At mm-wave frequencies, misalignment between the aperture and patch can lead to significant performance degradation.
- **Fabrication Complexity:** Requires multilayer PCB processes or hybrid fabrication, which can increase cost and manufacturing time.
- **Optimization Trade-offs:** The size and position of the aperture must be carefully optimized to balance coupling strength and minimize unwanted radiation.

3.3.4 Applications

Aperture-coupled antennas are widely employed in advanced radar systems, satellite communication, and high-speed mm-wave communication links. Their ability to support high gain, wide bandwidth, and isolation makes them suitable for performance-critical applications.

3.4 SIW Antennas

SIW technology has become a popular platform for implementing compact, low-loss, and cost-effective guided wave structures, especially useful at mm-wave frequencies. Unlike traditional waveguides made from metal tubes, SIW structures can be fabricated using standard PCB techniques, making them highly suitable for integration with planar circuits and systems.

3.4.1 What Is SIW?

SIW essentially mimics a conventional rectangular waveguide but in a planar form. The structure consists of two parallel rows of metallic vias embedded in a dielectric substrate, sandwiched between two metal layers. These vias act like sidewalls of a waveguide and confine the electromagnetic fields within the dielectric medium.

The SIW supports transverse electric (TE) modes similar to those in a conventional waveguide. The dominant mode in most SIW structures is the TE_{10} mode. The equivalent width W_{eq} of an SIW is calculated to ensure it behaves similarly to a metallic waveguide:

$$W_{eq} = W_{SIW} - \frac{d^2}{0.95p}, \tag{3.9}$$

where:

- W_{SIW} is the center-to-center spacing between the rows of vias,
- d is the diameter of the metallic vias,
- p is the pitch between the vias.

To minimize leakage losses and ensure wave confinement, the following conditions are typically maintained:

$$p < 2d \quad \text{and} \quad d < \frac{\lambda_g}{5}. \tag{3.10}$$

Here, λ_g is the guided wavelength inside the dielectric.

3.4.2 Why SIW Is Great for mm-Wave?

At mm-wave frequencies (24 GHz and above), conventional waveguides offer excellent performance but are bulky and expensive to fabricate. On the other hand, microstrip lines are compact but suffer from high radiation and conductor losses. SIW provides a sweet spot: it offers low loss performance closer to waveguides, with the manufacturability of planar circuits.

Some standout advantages include:

- **Compactness:** Planar configuration makes SIW easy to integrate with other microwave and digital components.
- **Low Loss:** Less conductor and radiation loss compared to microstrip at high frequencies.
- **Fabrication Friendly:** Can be built using standard PCB processes, reducing cost and complexity.
- **High Power Handling:** Better than microstrip lines, which is critical in many mm-wave systems.

3.4.3 SIW Antenna Types

Many antenna configurations can be built using SIW technology, depending on the application and performance requirements. Some common types include:

3.4.3.1 SIW Cavity-backed Patch Antennas

In this design, the patch radiator is placed over an SIW cavity. The cavity helps confine the fields and improves bandwidth and radiation efficiency.

3.4.3.2 SIW Slot Antennas

Slots are etched on the top metal layer of the SIW. These slots can be resonant or non-resonant and are carefully placed and shaped to achieve desired radiation characteristics.

3.4.3.3 Tapered SIW Horn Antennas

By gradually increasing the SIW width, a horn-like radiation pattern can be achieved in planar form. These antennas offer high gain and directional radiation.

3.4.3.4 Array Antennas

SIW configuration is particularly advantageous when designing arrays. The feed network can also be implemented using SIW, leading to a highly integrated solution.

3.4.4 Design Considerations

When designing SIW antennas, several parameters require careful attention:

- **Dielectric Constant (ϵ_r):** Affects size and impedance. High ϵ_r leads to compact designs but may reduce bandwidth.
- **Substrate Thickness:** Should be chosen based on the desired mode, bandwidth, and radiation efficiency.
- **Via Design:** Vias should be closely spaced to avoid leakage and ensure good wave confinement.
- **Transition Design:** Transitions from microstrip to SIW must be smooth to avoid reflections and impedance mismatches.

A commonly used formula for the cutoff frequency of TE_{10} mode in SIW is:

$$f_c = \frac{c}{2W_{\text{eq}}\sqrt{\epsilon_r}}, \tag{3.11}$$

where c is the speed of light in vacuum.

3.4.5 Applications of SIW Antennas

SIW antennas are found in numerous applications such as:

- 5G base stations
- Automotive radar
- Satellite communication
- mm-Wave imaging systems
- Wireless backhaul networks

SIW antennas bridge the gap between traditional waveguides and planar microwave technologies. They offer high performance at mm-wave frequencies while being compact, cost-effective, and easy to manufacture. For engineers and researchers aiming to design modern high-frequency systems, SIW antennas are a practical and promising solution.

3.4.6 Design Example

The authors of Naseri et al. (2022) have designed an integrated microwave and mm-wave antenna based on SIW technology, as shown in Figure 3.9. It contains four ports, with each port operating at a distinct frequency band. The antenna resonates at 5.4 GHz when Port 1 is active, at 4.8 GHz when Port 2 is active, at 28 GHz when Port 3 is active, and at 30 GHz when Port 4 is active (Figure 3.10). A full-mode SIW cavity is considered and then halved to design the HMSIW. The microwave frequency bands utilize the HMSIW cavity, while the mm-wave frequency bands use a portion of the same cavity. Since all antennas are in an integrated form, sufficient isolation between them is important. In this design, a reasonable isolation of more than 20 dB is maintained, as shown in Figure 3.11. The microwave frequencies exhibit gains of 5.4 dBi at 4.8 GHz and 5.2 dBi at 5.4 GHz. At the mm-wave frequency bands, the antenna achieves gains of 8 dBi at 28 GHz and 8.7 dBi at 30 GHz.

One of the key advantages of multiplexing antennas is that each operating frequency can be independently controlled. In other words, tuning one frequency does not significantly affect the others. To demonstrate this behavior, various design parameters—such as slot lengths, the number of vias, and the values of

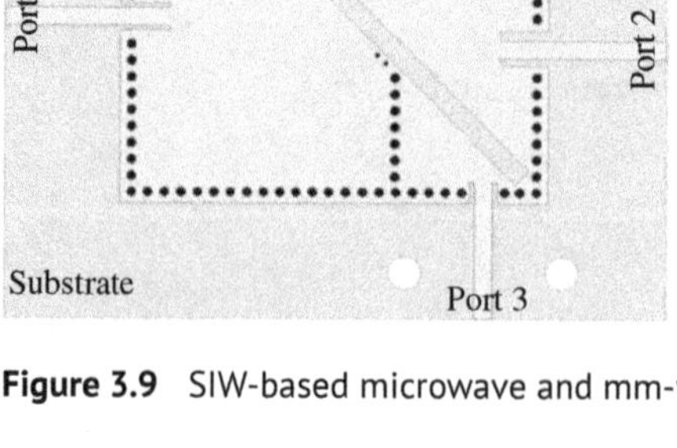

Figure 3.9 SIW-based microwave and mm-wave antenna configuration.

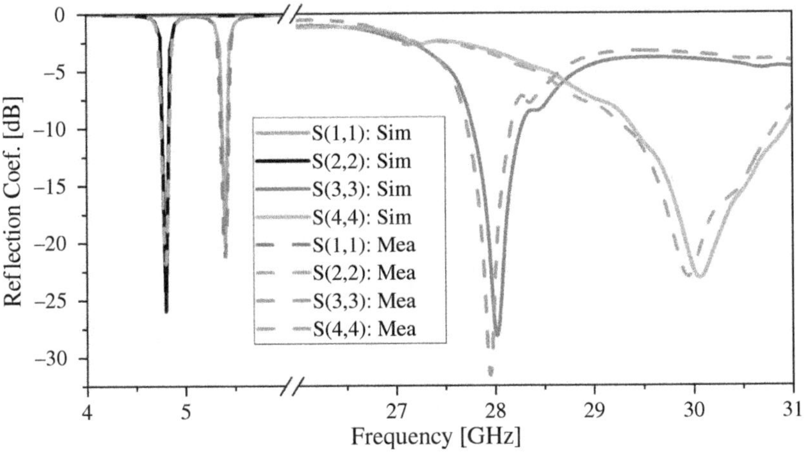

Figure 3.10 Reflection coefficient of the antenna.

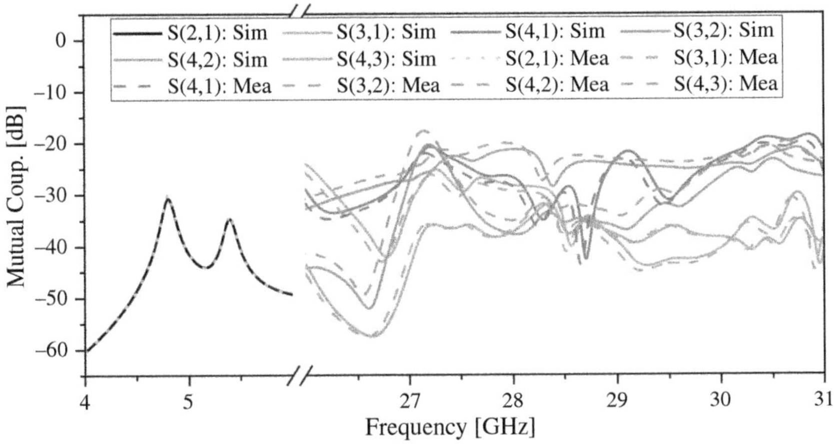

Figure 3.11 Mutual coupling between the ports of the antenna.

parameters G and Q—were systematically varied, and the corresponding effects on the antenna's response were observed.

As illustrated in Figure 3.12, increasing the length of the slot causes the corresponding resonant frequency to shift toward lower values. For example, when the slot length is 3.8 mm, the resonance occurs at 4.8 GHz. Increasing the length to 4.8 mm shifts the frequency to 4.6 GHz, while the other frequency bands remain largely unaffected.

A similar trend is observed when adjusting the number of vias (Figure 3.13). Adding more vias reduces the effective radiating area, which in turn increases the

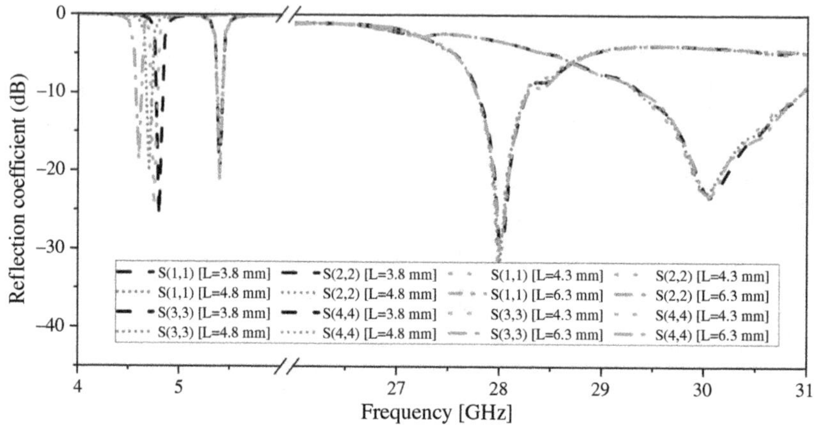

Figure 3.12 Reflection coefficient by varying parameter "L".

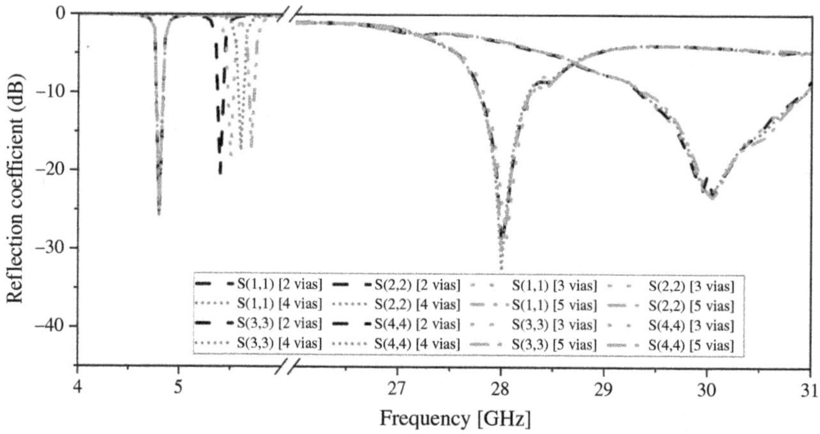

Figure 3.13 Reflection coefficient by varying the number of vias.

resonant frequency. Although results for variations in via spacing and diameter are not shown here for brevity, it was observed that reducing the spacing or diameter of the vias around the cavity opening tends to raise the resonance frequency.

3.5 Millimeter-wave Slot Antennas

Slot antennas have gained significant traction in the mm-wave spectrum due to their low profile, planar structure, mechanical robustness, and ease of integration with planar circuits such as SIW or microstrip lines. These antennas are particularly advantageous in high-frequency applications where compactness and

Figure 3.14 Configuration of mm-wave SIW-based slot antenna.

manufacturability are critical. An example of a slot antenna based on SIW technology is shown in Figure 3.14.

3.5.1 Principle of Operation

A slot antenna typically consists of a narrow slot cut into a conductive surface, which can be excited using a variety of feed mechanisms—most commonly via microstrip lines or waveguides. According to Babinet's principle and the complementary antenna theory, the radiation pattern of a slot antenna is similar to that of its complementary dipole, rotated by 90 degrees and with orthogonal polarization.

The most basic type, a rectangular slot, is usually designed to resonate when the slot length L_s is approximately half of the guided wavelength λ_g, i.e.,

$$L_s \approx \frac{\lambda_g}{2}. \tag{3.12}$$

Here, the guided wavelength λ_g in a dielectric substrate is given by:

$$\lambda_g = \frac{\lambda_0}{\sqrt{\varepsilon_{\text{eff}}}}, \tag{3.13}$$

where:

- $\lambda_0 = \frac{c}{f}$ is the free-space wavelength,
- c is the speed of light,

- f is the operating frequency,
- ε_{eff} is the effective dielectric constant of the substrate.

3.5.2 Slot Width and Bandwidth Considerations

The width of the slot W_s has a substantial impact on the antenna's impedance bandwidth. A wider slot generally provides a broader bandwidth but may compromise radiation efficiency. The input impedance of a slot antenna depends on the slot width, the substrate properties, and the feed mechanism. An approximate relation for the slot impedance at resonance is given by:

$$Z_{\text{in}} \propto \frac{1}{W_s}. \tag{3.14}$$

This suggests that a narrower slot leads to a higher impedance, which can be useful in impedance matching when integrated with high-impedance feed lines.

3.5.3 Feeding Mechanisms

At mm-wave frequencies, accurate excitation becomes critical due to increased losses and fabrication tolerances. Common feeding techniques include:

- Microstrip-line coupled slots
- CPW fed slots
- SIW fed slots

The most efficient feed method depends on the targeted bandwidth, fabrication complexity, and integration needs with other components.

3.5.4 Applications and Advantages

Millimeter-wave slot antennas are widely used in:

- On-chip and in-package antennas for 5G and 6G systems
- Radar and imaging systems
- Body-centric and wearable devices

The key advantages include:

- High radiation efficiency in planar form
- Compatibility with monolithic microwave integrated circuits (MMICs)
- Suitability for conformal and array designs

3.5.5 Design Note for Arrays

For high-gain applications, linear or planar arrays of slot antennas are often used. The element spacing d in such arrays is typically kept below:

$$d \leq \frac{\lambda_0}{2}, \tag{3.15}$$

to avoid grating lobes in the radiation pattern. Mutual coupling effects between closely spaced slots must also be carefully considered during array design, especially at mm-wave frequencies.

3.6 Conclusion

In this chapter, a closer look has been taken at the role of planar antennas in mm-wave systems, with a particular focus on microstrip patch and slot antennas. These compact and flexible antenna types are becoming increasingly important as wireless technologies push into higher frequency bands.

Microstrip patch antennas are especially popular thanks to their straightforward design and the ease with which they can be tuned by adjusting physical dimensions or material properties. Slot antennas, meanwhile, offer a slim profile and are known for their wide bandwidth and adaptability to different feed mechanisms: traits that make them ideal for emerging applications like 5G and radar systems.

Of course, designing antennas at these high frequencies comes with its own set of challenges. Engineers must carefully balance theoretical design with real-world constraints like fabrication tolerances and material performance. Each type of antenna (patch or slot) has its strengths and trade-offs, and researchers continue to refine both to meet growing demands.

Looking ahead, developments in materials, fabrication technologies, and design techniques are expected to open up even more possibilities. These advances will help address the rising need for faster, more reliable communication and sensing systems, keeping pace with the rapid evolution of wireless technologies.

References

Elagandula Aparna, Gopi Ram, and G Arun Kumar. 5G mm-wave technology: Innovative design of integrating mm-wave wideband antenna with a compact CP microwave antenna for diverse applications. *IEEE Access*, 12:56633–56641, 2024.

Constantine A Balanis. *Antenna Theory: Analysis and Design*. John Wiley & Sons, 2016.

Niamat Hussain and Nam Kim. Integrated microwave and mm-wave mimo antenna module with 360° pattern diversity for 5G internet of things. *IEEE Internet of Things Journal*, 9(24):24777–24789, 2022. https://doi.org/10.1109/JIOT.2022.3194676.

Jun Li, Hongli He, Shijie Xiong, Guangyin Feng, and Yanjie Wang. A novel high-gain antipodal vivaldi antenna for 5G mm-wave applications. *AEU—International Journal of Electronics and Communications*, 177:155160, 2024.

Hassan Naseri, Peyman PourMohammadi, Amjad Iqbal, Ahmed A. Kishk, and Tayeb A. Denidni. SIW-based self-quadruplexing antenna for microwave and mm-wave frequencies. *IEEE Antennas and Wireless Propagation Letters*, 21(7):1482–1486, 2022. https://doi.org/10.1109/LAWP.2022.3172007.

Arun Raj and Durbadal Mandal. Design and experimental analysis of fractal antennae with ground-defected structure for expected 6G and 5G mm-wave communication and wireless applications. *Transactions on Emerging Telecommunications Technologies*, 35(1):e4900, 2024.

Rakesh N Tiwari, Deepti Sharma, Prabhakar Singh, and Pradeep Kumar. A flexible dual-band 4 × 4 MIMO antenna for 5G mm-wave 28/38 GHz wearable applications. *Scientific Reports*, 14(1):14324, 2024.

4

Millimeter-wave Dielectric Resonator Antennas

4.1 Introduction

Typically, the frequency bands for millimeter waves lie between 30 and 300 GHz and have gained prominence in recent times for many applications. These bands could be good candidates for advanced wireless communication systems, such as fifth-generation (5G) and sixth-generation (6G) networks, which are being proposed/will be proposed in the future, automotive radar systems, high-speed wireless connectivity, and sensing/imaging techniques. The major benefit of mm-wave technology is that it provides a wide bandwidth and hence offers high speeds and lower latencies. However, operating at such high frequencies also introduces several challenges, especially in antenna design. At millimeter-wave frequencies, metallic losses rise, fabrication tolerances become very critical, and conventional antennas mostly demonstrate reduced efficiency.

4.2 Dielectric Resonator Antennas

Dielectric resonator antennas (DRAs) have become a strong alternative to conventional metallic antennas for mm-wave applications. A typical DRA is made of a dielectric material of relatively high permittivity and is usually placed on a ground plane (Alanazi, 2023; Keyrouz and Caratelli, 2016). Various feeding techniques are possible, including microstrip lines, coplanar waveguide (CPW) lines, probes, or aperture coupling. Unlike microstrip patch antennas, DRAs do not rely on surface currents flowing on metal conductors to produce radiation. Instead, the electromagnetic fields confined within the dielectric resonator are responsible for radiation. Some common modes excited within the DRA are shown in Figure 4.1. This basic underlying principle leads to a number of advantages for DRAs, especially at higher frequencies.

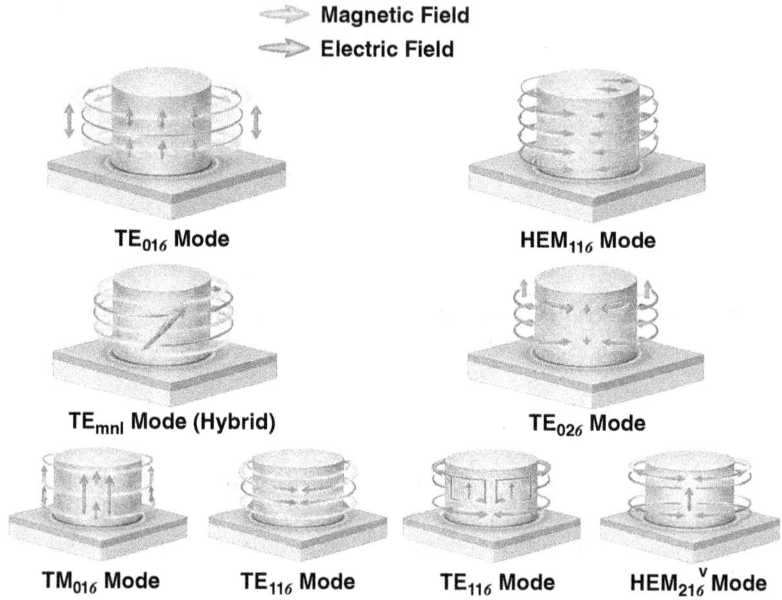

Figure 4.1 Illustration of dominant resonant modes in a cylindrical DRA.

The use of dielectric resonators was, until recently, confined to microwave components, such as filters and oscillators. It was not until later that it was realized that these resonators can efficiently radiate when properly excited. Since then, DRAs have been under extensive study and development for antenna applications. As the dielectric material and fabrication technologies improve further, DRAs become even more practical and attractive for compact and high-performance mm-wave systems (Khattak et al., 2020; Zhang et al., 2019).

4.3 Advantages of DRAs at mm-Wave Frequencies

At millimeter-wave frequencies, DRAs offer several significant advantages. One of their most important advantages is the low conduction loss at mm-wave frequencies. Conventionally, radiation in metallic antennas depends on surface currents flowing on conductors. At mm-wave frequencies, the skin depth becomes very small, increasing ohmic losses and reducing antenna efficiency. DRAs, however, are mainly made of dielectric material and do not need metal radiators. In this case, conduction losses are minimal, and radiation efficiency is higher. This property makes DRAs well-suited for mm-wave applications, where efficiency is a key requirement (Nezhad-Ahmadi et al., 2010).

Another important advantage of DRAs is their wide impedance bandwidth. Most mm-wave systems necessitate a wide bandwidth for high data rates and reliable communication. Comparatively, wider bandwidths can be achieved with DRAs than with the traditional microstrip patch antennas. For a DRA, bandwidth can easily be controlled by choosing proper dielectric constants, resonator dimensions, and excitation methods. Further bandwidth enhancement is normally achieved using lower permittivity material and higher-order modes of operation. This flexibility in design is very useful in practical mm-wave antenna systems.

DRAs also exhibit high radiation efficiency for small antenna sizes. At mm-wave frequencies, antennas are naturally compact because of the short wavelength. In general, however, size reduction in metallic antennas suffers from reduced efficiency. Because DRAs radiate through volume modes instead of surface currents, they could maintain a good performance even at small sizes, and hence, they are suited for compact mm-wave devices like mobile terminals, sensors, and integrated modules.

Another great advantage of DRAs is their design flexibility. They can be fabricated into several shapes that include, but are not limited to, rectangular, cylindrical, and hemispherical, among other complex geometries. Each geometry supports different resonant modes and radiation characteristics. This enables variations in bandwidth, polarization, and radiation pattern to be selected according to specific requirements by choosing a suitable geometry. This becomes of great benefit in mm-wave applications where space is limited, and integration is important.

DRAs are also easy to integrate with planar RF circuits. They can be excited using common feeding techniques such as microstrip lines, CPWs, coaxial probes, or slot apertures. All these techniques make DRAs compatible with standard printed circuit board technologies. For mm-wave systems where antennas are often integrated with front-end circuits, such compatibility simplifies the overall system design and reduces fabrication complexity.

One of the benefits of DRAs is being able to support multiple resonant modes. It is possible to excite multiple resonant modes using a single dielectric resonator. This technique has been used to design dual-band, multi-band, or wideband antennas. It is very beneficial in mm-wave communication, where it is possible to have multiple frequency bands or communicate/sense using a single antenna (Ma et al., 2022; Yang et al., 2020).

DRAs are also known to provide stable radiation patterns and gain over a broad frequency band. On a comparative basis, DRAs display more stable gain variations and stable radiation patterns as opposed to certain metallic antennas. Such stable performance is very essential in mm-wave links where the signal integrity may vary with the radiation pattern.

Based on the perspective of fabrication, DRAs have simpler structures with no metal patterns. When it comes to mm-wave frequency bands, effective fabrication is very important. There would be fewer opportunities for faulty manufacturing in simpler structures. There has been significant progress in the ceramic material technology that offers methods for better dielectric resonator designs.

Finally, DRAs are able to manage higher levels of power. Since there are no thin metallic lines that tend to exhibit heating effects, it becomes feasible for the DRAs to manage a rather higher level of transmitted power without compromising their performance (Meher et al., 2020; Pan et al., 2011, 2019).

4.4 Dielectric Materials for DRAs

The heart of DRAs comprises dielectric materials and their shapes. The electromagnetic characteristics of a DRA are mostly a function of the electrical properties of the chosen dielectric material constituting that particular resonator design. While operating in mm-wave frequency bands, this aspect becomes even more fundamental, as losses and errors during fabrication tend to influence antenna efficiency and reliability. Careful selection of the dielectric material is, therefore, necessary in achieving good performance for the designs of DRAs at mm-wave frequencies.

The main parameter of a dielectric material is the relative permittivity, usually denoted by ε_r. The value of this parameter defines the resonator size and thus the resonant frequency of the antenna. High permittivity allows reduction of the resonator size, making it useful for compact mm-wave systems. At the same time, high values of permittivity give strong field confinement inside the dielectric; this can reduce the radiation efficiency and bandwidth. For lower values of permittivity, larger resonators are obtained; however, the obtained bandwidth is generally wider, and the radiation performance is better. In the design, therefore, one has to make a compromise between the size, bandwidth, and efficiency while choosing ε_r.

The other important characteristic of dielectric materials is the loss tangent, which is often denoted by $\tan\delta$. Loss tangent gives the dielectric losses in the material and, hence, directly influences the antenna efficiency. Even a small loss in the dielectric at mm-wave frequencies causes considerable performance deterioration. Hence, materials with a very low loss tangent are preferred for mm-wave DRAs. High-quality ceramic materials are widely used due to their low loss, stable permittivity, and good temperature stability. Low loss tangent ensures that most of the input power will be radiated instead of being dissipated as heat within the dielectric.

The stability of materials within mm-wave applications is another important aspect. It is expected that the dielectric constant and loss tangent will be invariant for a wide frequency range and within a great span of environmental conditions, such as temperature and humidity. Variations in these parameters will shift the resonant frequency and affect impedance matching conditions. In practice, especially for automotive and outdoor applications of mm-wave systems, stable dielectric properties are required to maintain consistent antenna performance.

Various kinds of dielectric materials have been reported for DRA applications. Ceramic materials are the most common choice because of their high permittivity and low-loss characteristics. In addition, alumina-based ceramics, titanium dioxide-based materials, or other engineered ceramic compounds are frequently used in mm-wave DRAs. These materials provide good mechanical strength and can be fabricated with high precision, which is very important at mm-wave frequencies, where small dimensional errors can lead to large frequency shifts.

Other than traditional ceramics, some DRA designs use polymer-based dielectric materials. These tend to have lower permittivity and losses compared to ceramics. However, they offer the benefits of being low-cost, easy to process, and fully compatible with the regular PCB technology processes. The application of dielectric polymers may therefore be applicable in situations where ultra-compact size is not the primary requirement.

There have been several breakthroughs in material sciences that have given new possibilities to DRA design. Modern processing methods of additive manufacturing and 3D printing allow the design of dielectric resonators with complex geometry and permittivity variation. Such approaches can be used to optimize the bandwidth, radiation patterns, or multiband performance. Though these approaches have not yet reached full maturity, they possess enormous potential for future mm-wave DRAs.

Another thing worth noticing is how results may be affected by interaction between the dielectric resonator and either the support substrate or ground plane, caused by air gaps, layers of bonding, and even substrate roughness, all of which can impact permittivity and hence the overall performance of the antenna. In particular, at millimeter-wave frequencies, such differences can be even more pronounced and need careful consideration during design and fabrication.

In short, the importance of the dielectric materials in the successful design of the mm-wave DRA antennas cannot be overemphasized. The value of the relative permittivity, tan delta, and stability of the dielectric materials have a major influence on the size, efficiency, and bandwidth of the designed antennas. By properly choosing and designing the materials, the designers are able to make full use of the potential of the DRAs and meet the most challenging requirements of the modern mm-wave systems. The summary of key material considerations for mm-wave DRAs is shown in Figure 4.2.

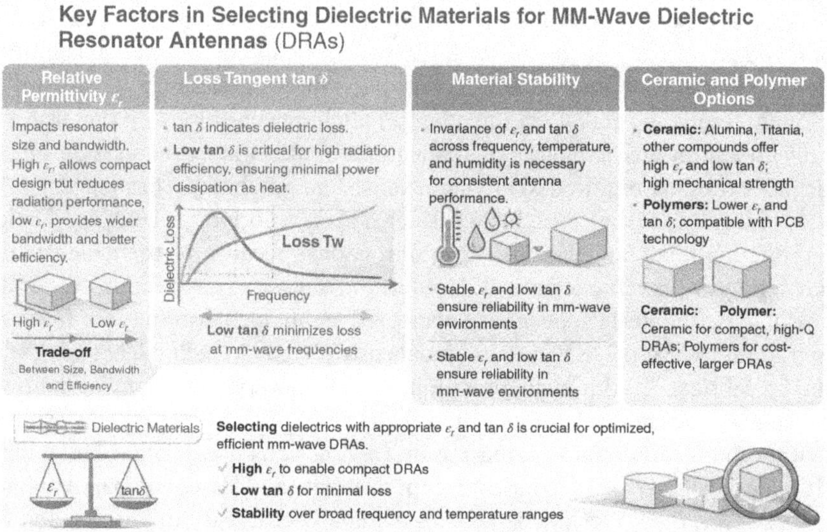

Figure 4.2 Summary of key material considerations for mm-wave dielectric resonator antennas.

4.5 Conductor and Dielectric Losses

Losses are a very important factor in antenna performance, especially at mm-wave frequencies. Generally speaking, losses of antennas can be divided into conductor losses and dielectric losses. Both types of losses reduce the radiation efficiency and, therefore, limit the overall performance of the antenna.

The most important mechanism responsible for conductor loss is the finite conductivity of metals. At millimeter-wave frequencies, the skin depth is very small, and the current flows only on the thin layer of the conductor surface. This makes the effective resistance of the metal higher and causes higher power dissipation. As a result, conventional metallic antennas such as the microstrip patch antenna often have low efficiency at millimeter wave frequencies.

Loss inside the dielectric typically occurs because of energy dissipation during the application of an electromagnetic field to the material. It is normally described by the loss tangent of that material. At mm-wave frequencies, the dielectric loss increases, and hence even small values of losses can degrade the performance of the antenna. For this reason, low-loss dielectric materials are always favorable, particularly in high-frequency applications.

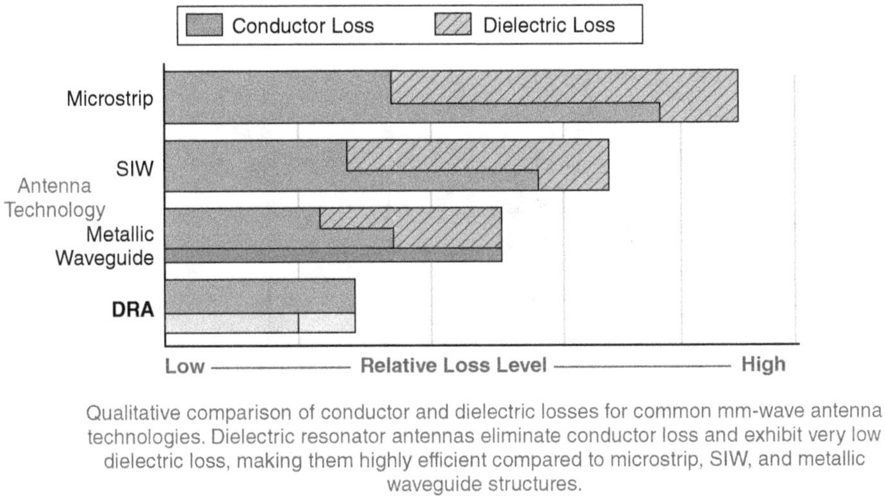

Figure 4.3 Qualitative comparison of conductor and dielectric losses in common mm-wave antenna technologies.

DRAs do not rely on surface currents on metal radiators and hence have an advantage in this aspect, with a significant reduction of conductor losses. However, the dielectric losses cannot be eliminated but depend on the quality of the dielectric material used. mm-Wave DRAs have significantly improved overall efficiency as low-loss-tangent and stable properties materials are selected. The qualitative comparison of conductor and dielectric losses in common mm-wave antenna technologies is shown in Figure 4.3.

4.6 Feeding Techniques

In the millimeter-wave DRA design process, several feeding techniques can be employed, as described in the following sections.

4.6.1 Coaxial Probe Feeding Technique

Coaxial probe feeding is a straightforward and widely used excitation technique for DRAs. This excitation technique consists of the inner conductor of the coaxial cable connected to the dielectric resonator, while the outer conductor is connected to the ground plane. The probe penetrates through the ground plane into the dielectric resonator, allowing energy to be directly coupled to the antenna. The illustration of the coaxial probe feeding technique is shown in Figure 4.4.

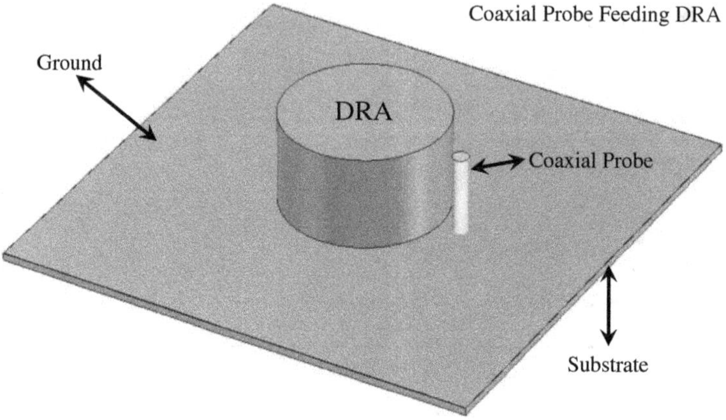

Figure 4.4 Illustration of coaxial probe feeding technique for circular DRA.

One merit of the coaxial probe feed is that it has a simple structure and can be easily designed without complicated fabrication steps. By adjusting the position and length of the probe inside the dielectric resonator, good impedance matching is easily achieved. This flexibility makes it well-suited for both single-band and multi-band DRA designs (Junker et al., 1994; Kishk et al., 2003; Leung et al., 1993).

Therefore, careful design of the probe is required at mm-wave frequencies. Small length or position changes in the probe are enough to cause noticeable shifts in resonant frequency and impedance matching. Another important issue of the probe is unwanted inductance that may be introduced, thus affecting bandwidth. Despite these constraints, the coaxial probe feeding technique still remains widely used because it is easy to implement and can provide good performance when properly designed.

Overall, the coaxial probe feeding technique is a very effective, simple method of exciting DRAs for most experimental studies and practical millimeter wave antenna prototypes.

4.6.2 Electromagnetically Coupled DRAs

Electromagnetic coupling remains one of the most used techniques for exciting DRAs without making an electrical contact with the resonator. In this methodology, the energy is coupled from a feeding structure to the dielectric resonator through electromagnetic fields. Common feeding topologies include microstrip lines, CPWs, or slots etched in the ground plane. The dielectric resonator is placed close to the feed and is coupled through the near-field interaction. The illustration of the electromagnetically coupled DRA technique is shown in Figure 4.5.

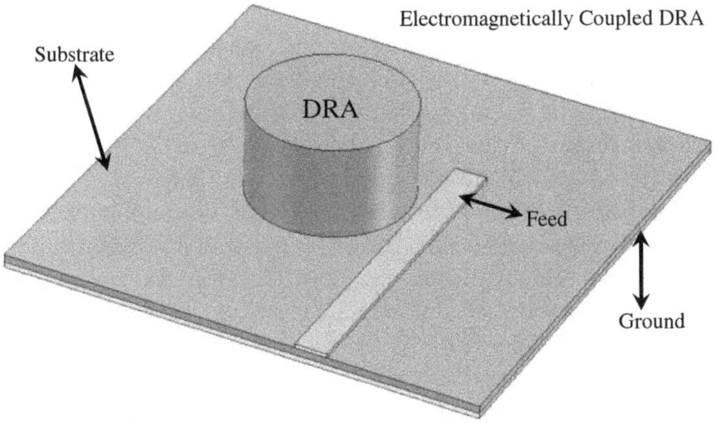

Figure 4.5 Illustration of electromagnetically coupled circular DRA.

One of the major advantages of electromagnetic coupling is improved isolation between the feed and the radiating element. Because there is no direct connection, the feeding network has less influence on the radiation characteristics of the antenna. This would enable cleaner radiation patterns and better impedance stability, especially at mm-wave frequencies where direct contacts may introduce unwanted parasitic effects (Kishk et al., 2007; Soren and Poddar, 2014).

Furthermore, the degree of coupling in the electromagnetically coupled DRA can be controlled. Using the size of the slot, the distance between the feeding point and the resonator, or even the positioning of the dielectric block with respect to the resonator, the coupling degree can easily be changed. By properly tuning these parameters, a wide bandwidth and good impedance matching can be achieved. This makes the technique suitable for wideband and multi-band mm-wave applications.

The other advantage of such feeding is ease of fabrication and integration: since the feed is normally planar, it can be easily integrated with standard PCB technology. That is very useful for mm-wave systems where antennas and RF circuits normally share a common substrate. Overall, an electromagnetically coupled DRA can provide a practical as well as effective solution for achieving a high-performance millimeter-wave antenna featuring reduced losses and stable radiation behavior.

4.6.3 Aperture-coupled DRAs

The aperture-coupled feeding technique is a popular method for exciting DRAs, especially at mm-wave frequencies. In this approach, energy is coupled from a microstrip line to the dielectric resonator through a small slot or aperture etched in the ground plane. The feed line is placed on one side of the ground plane, while the dielectric resonator is positioned on the other side, directly above the aperture. The electromagnetic fields passing through the slot excite the resonant modes of the dielectric resonator (Iqbal et al., 2019; Naeem et al., 2017). The illustration of the aperture-coupled DRA technique is shown in Figure 4.6.

Aperture coupling has one major advantage concerning the coupling between the feed and the radiating element, which provides good isolation between the feed and the radiating element. This is because introducing a ground plane between the feed and the dielectric resonator helps to reduce parasitic radiation of signals through the feed line. This is particularly beneficial at mm-wave frequencies, where coupling effects are more pronounced.

Aperture-coupled DRAs also offer design flexibility. The degree of coupling can be varied by optimizing the size, shape, and location of the aperture, as well as varying the alignment of the dielectric resonator. Through this technique, a high degree of impedance matching and bandwidth can be realized. Thus, aperture coupling is useful for wideband and multi-band mm-wave designs.

Another advantage of this method is its compatibility with planar fabrication technology. The feeding structure can be realized with a standard PCB technology and easily integrated with the RF circuits. In spite of having a slightly more complex structure than the direct probe feed method, the aperture-coupled DRAs are superior in their performance and radiation characteristics, which make them very effective in the mm-wave antenna systems.

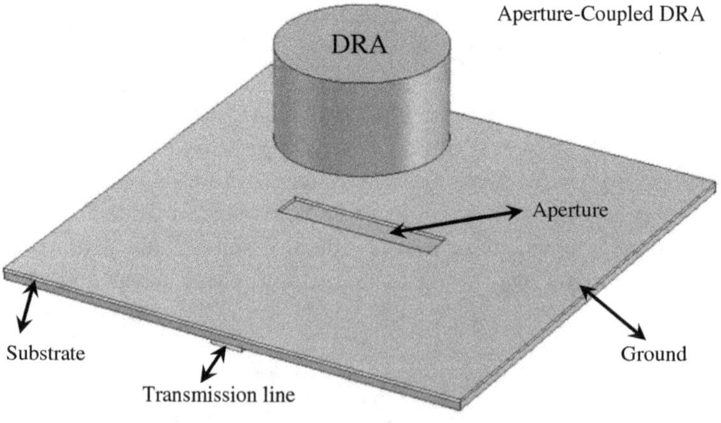

Figure 4.6 Illustration of aperture-coupled circular DRA.

4.6.4 CPW-fed DRAs

The CPW feeding method is a very efficient feeding technique for the excitation of DRAs, especially in mm-wave devices. In this method, the signal trace and ground traces are located in the same plane on the substrate. The dielectric resonator is placed near the CPW transmission line or immediately above it such that the electromagnetic wave is coupled from the transmission line due to its fringing fields (Al Salameh et al., 2002; Deng et al., 2001; Gao et al., 2011). The illustration of the aperture-coupled DRA technique is shown in Figure 4.7.

One of the most significant benefits of CPW-fed DRAs is their simpler structure. The presence of the signal trace and ground plane on the same side of the substrate results in simpler manufacturing, which is especially preferable for the mm-wave frequency band, as manufacturing processes such as alignment of layers can be challenging.

In terms of performance and flexibility, CPW-fed DRAs guarantee fine impedance matching and a broad bandwidth. However, the coupling factor can be varied according to the distance between the signal trace and ground, the width of the signal trace on the CPW line, and the location of the dielectric resonator. This flexibility enables the optimal antenna design for various mm-wave bands.

In terms of integration and practical benefits, another major advantage of CPW feeding is its compatibility with active and passive components of RF circuits. CPW transmission lines can be found in various mm-wave integrated circuits. Hence, these lines can be considered suitable for integration with an antenna. Moreover, CPW-fed DRAs can provide stable radiation patterns and controlled spurious radiation, though they might lack these features when using other feeding technologies. Thus, CPW-fed DRAs prove to be an optimal solution for mm-wave antennas.

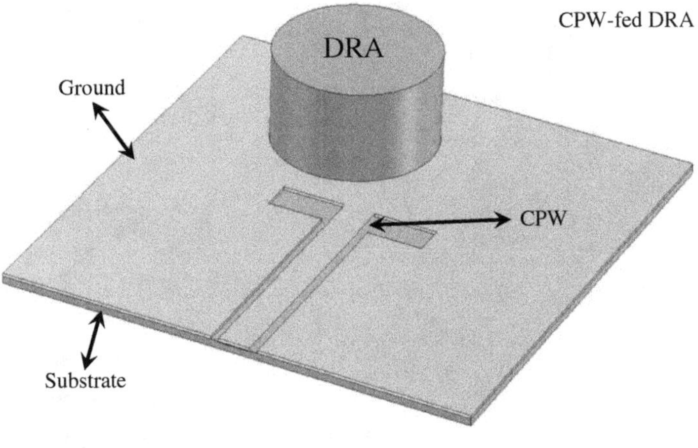

Figure 4.7 Illustration of CPW-fed circular DRA.

4.7 Applications of DRAs

DRAs are widely employed in advanced wireless communication systems owing to their efficient performance and stable characteristics. The DRAs are very suitable for working with high-frequency communications. Along with the development of wireless communications, the employment of DRAs in both practical and research work has naturally expanded. The visual representation of different applications of DRAs is shown in Figure 4.8.

4.7.1 Wireless Communication Systems

One of the major applications of DRAs is in mm-wave wireless communication. For example, 5G and even 6G communication systems will operate at mm-wave frequency ranges and will require antennas that should have wide-band operation along with efficiency and compact size. DRAs fulfill such conditions and are easily integrated with RF front-end circuits to support high data rate communication in base stations, small cells, and user terminals.

4.7.2 Automotive Radar Systems

DRAs are also widely used in automotive radars. Nowadays, cars employ the use of radars in order to perform tasks like collision avoidance, adaptive cruise control, and blind spot detection. The radars used in these tasks operate in the mm-wave

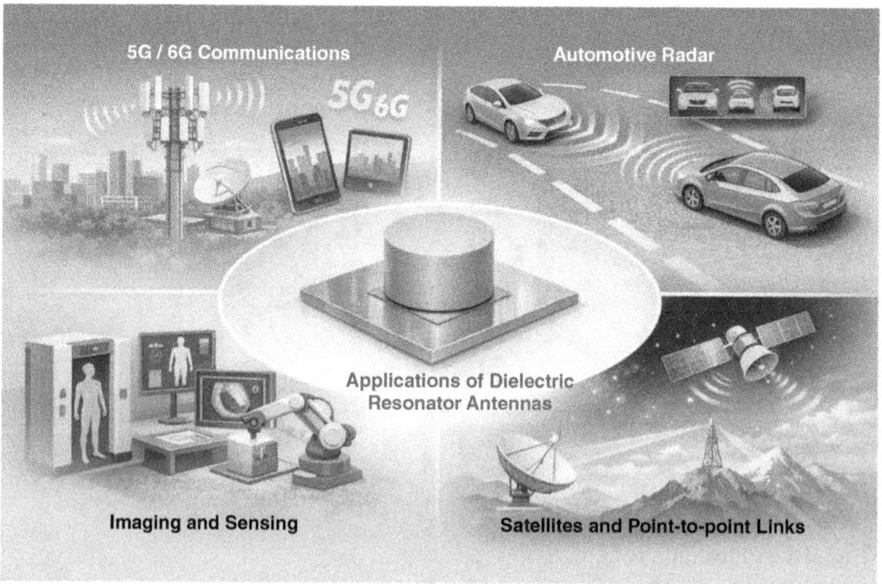

Figure 4.8 Visual representation of different applications of DRAs.

frequency band, for instance, at 77 GHz. DRAs are used in the mentioned application due to their advantages, which include stable radiation patterns, efficiency, and their ability to handle a lot of power.

4.7.3 Wireless Sensing and Imaging

Another significant use of DRAs is in wireless sensing and imaging. At the mm-wave range, antennas are utilized in security scanning, medical imaging, and material analysis. The low loss and satisfactory radiation properties of DRAs result in improved resolutions in imaging and the design of accurate imaging devices. Due to the compact nature, DRAs facilitate integration into imaging systems for both portable and fixed devices.

4.7.4 Satellite and Point-to-point Links

DRAs are also used in satellite and point-to-point communication links. These systems require antennas with high gain and stable performance over wide frequency ranges. DRAs can be used as single elements or as part of antenna arrays to meet these requirements. Their low loss and high efficiency make them suitable for long-distance and high-frequency communication links.

4.7.5 Research and Experimental Use

In addition, DRAs are increasingly used in research and experimental platforms. Due to their simple structure and flexible design, DRAs are often chosen for studying new antenna concepts, multi-band operation, and advanced feeding techniques. Overall, DRAs play an important role in many mm-wave applications and are expected to become even more important in future wireless and sensing technologies.

4.8 Conclusion

The chapter has presented a broad review of DRAs, with a focus on using mm-waves. The design fundamentals of DRAs, together with their advantages, materials, losses, and excitation techniques, have been addressed in a relevant manner in a bid to raise adequate knowledge on their efficiency in higher bands of frequencies. One of the advantages of using DRAs over other conventional metallic antennas is reduced losses, which occur in mm-waves.

This chapter has further focused on the significance of dielectric materials and the method of feeding that needs to be done while designing DRAs for effective impedance matching and better radiation characteristics. Various designs related to the method of feeding, such as probe-fed, electromagnetic-fed, aperture-fed, and

CPW-fed designs, have been discussed to identify the potential of such designs for various needs. The application areas of the DRA further reveal its significance and uses.

On a general note, DRAs represent a powerful and flexible technology for use in future generations of mm-wave communications and systems. As a result of improvements in materials and production processes, future high-frequency wireless and sensing applications will also be driven by DRAs.

References

MS Al Salameh, Yahia MM Antar, and Guy Seguin. Coplanar-waveguide-fed slot-coupled rectangular dielectric resonator antenna. *IEEE Transactions on Antennas and Propagation*, 50(10):1415–1419, 2002.

Meshari D Alanazi. A review of dielectric resonator antenna at mm-wave band. *Eng*, 4(1):843–856, 2023.

Sheng-Ming Deng, Tsung-Wen Chen, and Hsiao-Hsun Kan. A CPW-fed rectangular dielectric resonator antenna. In *APMC 2001. 2001 Asia-Pacific Microwave Conference (Cat. No. 01TH8577)*, vol. 2, pages 954–957. IEEE, 2001.

Yang Gao, Zhenghe Feng, and Li Zhang. Compact CPW-fed dielectric resonator antenna with dual polarization. *IEEE Antennas and Wireless Propagation Letters*, 10:544–547, 2011.

Amjad Iqbal, Amor Smida, Omar A Saraereh, Qais H Alsafasfeh, Nazih Khaddaj Mallat, and Byung Moo Lee. Cylindrical dielectric resonator antenna-based sensors for liquid chemical detection. *Sensors*, 19(5):1200, 2019.

G.P. Junker, A.A. Kishk, and A.W. Glisson. Input impedance of dielectric resonator antennas excited by a coaxial probe. *IEEE Transactions on Antennas and Propagation*, 42(7):960–966, 1994. https://doi.org/10.1109/8.299598.

Shady Keyrouz and Diego Caratelli. Dielectric resonator antennas: Basic concepts, design guidelines, and recent developments at millimeter-wave frequencies. *International Journal of Antennas and propagation*, 2016(1):6075680, 2016.

Muhammad Anab Khattak, Muhammad Irfan Khattak, Syed Muhammad Owais, Abbas Ali Khattak, and Asif Sultan. Design and analysis of millimeter wave dielectric resonator antenna for 5G wireless communication systems. *Progress In Electromagnetics Research C*, 98:239–255, 2020.

Ahmed A Kishk, Yahia MM Antar, and JL Volakis. Dielectric resonator antennas. *Antenna Engineering Handbook*, 4, 2007.

A.A. Kishk, Xiao Zhang, A.W. Glisson, and D. Kajfez. Numerical analysis of stacked dielectric resonator antennas excited by a coaxial probe for wideband applications. *IEEE Transactions on Antennas and Propagation*, 51(8):1996–2006, 2003. https://doi.org/10.1109/TAP.2003.814735.

K.W. Leung, K.M. Luk, K.Y.A. Lai, and D. Lin. Theory and experiment of a coaxial probe fed hemispherical dielectric resonator antenna. *IEEE Transactions on Antennas and Propagation*, 41(10):1390–1398, 1993. https://doi.org/10.1109/8.247779.

Chaojun Ma, Shao Yong Zheng, Yong Mei Pan, and Zhe Chen. Millimeter-wave fully integrated dielectric resonator antenna and its multi-beam application. *IEEE Transactions on Antennas and Propagation*, 70(8):6571–6580, 2022. https://doi.org/10.1109/TAP.2022.3161361.

Priya R Meher, Bikash R Behera, and Sanjeev K Mishra. Design and its state-of-the-art of different shaped dielectric resonator antennas at millimeter-wave frequency band. *International Journal of RF and Microwave Computer-Aided Engineering*, 30(7):e22221, 2020.

Umair Naeem, Amjad Iqbal, Muhammad Farhan Shafique, and Stéphane Bila. Efficient design methodology for a complex DRA-SIW filter-antenna subsystem. *International Journal of Antennas and Propagation*, 2017(1):6401810, 2017.

Mohammad-Reza Nezhad-Ahmadi, Mohammad Fakharzadeh, Behzad Biglarbegian, and Safieddin Safavi-Naeini. High-efficiency on-chip dielectric resonator antenna for mm-wave transceivers. *IEEE Transactions on Antennas and Propagation*, 58(10):3388–3392, 2010. https://doi.org/10.1109/TAP.2010.2055802.

Yong-Mei Pan, Kwok Wa Leung, and Kwai-Man Luk. Design of the millimeter-wave rectangular dielectric resonator antenna using a higher-order mode. *IEEE Transactions on Antennas and Propagation*, 59(8):2780–2788, 2011. https://doi.org/10.1109/TAP.2011.2158962.

Yong Mei Pan, Xin Qin, Yu Xiang Sun, Shao Yong Zheng. A simple decoupling method for 5G millimeter-wave MIMO dielectric resonator antennas. *IEEE Transactions on Antennas and Propagation*, 67(4):2224–2234, 2019. https://doi.org/10.1109/TAP.2019.2891456.

Dipali Soren and Dipak Poddar. Dielectric resonator antennas: Designs and advances. *Progress In Electromagnetics Research B*, 60:195–213, 2014.

Mei Di Yang,Yong-Mei Pan, Yu-Xiang Sun, Kwok-Wa Leung. Wideband circularly polarized substrate-integrated embedded dielectric resonator antenna for millimeter-wave applications. *IEEE Transactions on Antennas and Propagation*, 68(2):1145–1150, 2020. https://doi.org/10.1109/TAP.2019.2938629.

Yin Zhang, Jing-Ya Deng, Ming-Jie Li, Dongquan Sun, and Li-Xin Guo. A mimo dielectric resonator antenna with improved isolation for 5G mm-wave applications. *IEEE Antennas and Wireless Propagation Letters*, 18(4):747–751, 2019. https://doi.org/10.1109/LAWP.2019.2901961.

5

Substrate Integrated Waveguide Based mm-Wave Antennas

5.1 Introduction

Substrate integrated waveguide (SIW) technology has become a stable platform for millimeter-wave (mm-wave) circuits and antennas over the past two decades. It offers a practical middle ground between traditional metallic waveguides and planar transmission lines such as microstrip or coplanar waveguides. The motivation is straightforward: mm-wave systems require low loss, good confinement of fields, and affordable fabrication. Conventional waveguides provide excellent performance but are bulky and expensive to manufacture, while planar lines are easy to fabricate but suffer from high radiation loss and limited power handling at high frequencies. SIW was introduced to bridge this gap using standard printed-circuit processes, allowing waveguide-like behavior to be implemented directly on planar substrates (Li et al., 2020; Naseri et al., 2022; PourMohammadi et al., 2023).

An SIW is formed by placing two parallel rows of metallized via holes in a dielectric substrate and connecting the top and bottom metal layers. These vias emulate the sidewalls of a rectangular waveguide. A typical two-port SIW component is shown in Figure 5.1. Although the physical walls are not continuous metal plates, the periodic via fence is dense enough to confine fields in the dominant TE_{10} mode with low leakage. When the via diameter, spacing, and separation follow established empirical rules, the structure behaves like a conventional waveguide with predictable dispersion and impedance properties. This simple construction allows SIW circuits to be produced using standard multilayer PCB technology at low cost and with high yield (Althuwayb, 2021; Kumawat and Joshi, 2020).

At mm-wave frequencies, low loss becomes critical. Transmission lines such as microstrip and coplanar waveguide experience increasing conductor and dielectric losses as frequency increases. The unnecessary bends and discontinuities in transmission lines impact impedance matching also start radiation. These radiations are redundant and should be reduced while the impedance mismatch reduces the overall realized gain. These lines start radiation in case of large discontinuities

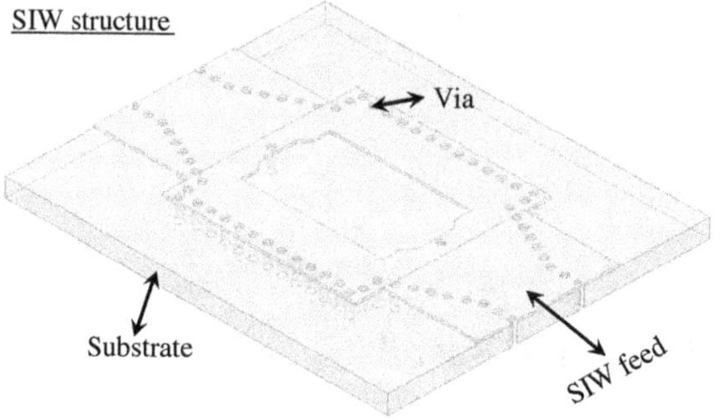

Figure 5.1 Example of two-port SIW structure.

in impedance sections. Moreover, bends in these lines also introduce unnecessary radiation. Such problems are more prominent in antennas working at high frequencies, such as the mm-wave frequency range. These problems are less prominent in SIW technology as fields are confined within such technology. The operational principles of SIW are the same as a metallic waveguide. In SIW, the currents are present on the inner side of the surface, which leads to low leakage and losses. Therefore, the losses in SIW are lower than in microstrip and CPW transmission lines (Pandey et al., 2024; Sun et al., 2020).

SIW technology has been widely adopted due to its several advantages. Many systems operating at the mm-wave frequency range use SIW for transmission lines as well as for active and passive RF circuits. These applications include automotive radars, mm-wave 5G, and mm-wave imaging. Attributes like SIW can be obtained from metallic waveguides. However, they are bulky, consume more space, and are costly. SIW technology is also suitable for system-in-substrate (SiS) applications where several RF components are designed on the same substrate.

Based on waveguide principles and theory, the behavior of SIW components can be judged and predicted. This is important for design consideration. The fundamental mode is cutoff frequency dependent, where the cutoff frequency has an indirect relationship with the width of the cavity. Unlike microstrip technology, SIW technology results are more reliable in analytical and EM wave simulations. Many classical RF components based on waveguides can be easily designed using SIW technology. RF and antenna engineers who are familiar with waveguide technology can design SIW-based RF circuits with less effort.

Compared to lower frequencies, mechanical tolerances are more prominent at high-frequency bands. As mentioned above, metallic waveguides work well compared to planar structures. However, precise printing and machining are required when these components operate above 60 GHz. Moreover, planar structures such as microstrip lines are very sensitive to any tolerance at such a high frequency band. In contrast, SIW losses can be controlled by precise via diameters and spacing between them. By placing them in an appropriate space and having proper dimensions, leakages can be avoided. Therefore, SIW is crucial for applications where the frequency band is higher than 60 GHz, such as 77 GHz automotive radars (Chen et al., 2022).

In this chapter, we first discuss single-band SIW antenna designs, followed by a presentation and discussion of multiband SIW antennas. We then continue with SIW antenna arrays. In addition, wideband SIW antennas are described and discussed. Finally, concluding remarks are provided in the last section.

5.2 Single-band SIW Antennas

Millimeter-wave antennas operating at only one frequency band are commonly designed using SIW technology. At the mm-wave frequency band, SIW circuits have lower losses compared to microstrip technology. Single-band resonances in SIW are normally generated by exciting the fundamental TE_{10} mode or TM_{10} mode of the cavity. Here, the cavity is formed by a series of vias arranged in a row. In most cases, the antenna radiates using a slot or edge of the cavity in the case of sub-modes (half-mode, quarter-mode, etc.). This section outlines the basic design steps, the rules that govern via placement, and the equations that set the resonant frequency of the SIW cavity (Alibakhshikenari et al., 2021). For example, an SIW-DRA integrated single-band antenna is shown in Figure 5.2 (Naeem et al., 2017). It consists of an SIW filter and DRA. The dielectric resonator antenna is excited using a rectangular slot. The SIW filter is fed using CPW lines in an SIW cavity. It can be observed that the integrated SIW-DRA antenna works at a single frequency band.

5.2.1 Waveguide Dimensions and Effective Width

The starting point is the effective width of the SIW, which represents the separation between the two equivalent metallic sidewalls. Because the walls are built from rows of metallized vias rather than solid metal, the actual physical width W_{SIW} differs from the effective width W_{eff}. For a typical design, the effective width follows the widely used expression:

$$W_{\text{eff}} = W_{\text{SIW}} - \frac{d^2}{0.95\,p}, \tag{5.1}$$

where d is the via diameter and p is the pitch between adjacent vias. This correction term accounts for the finite gap between vias. When the vias are placed close enough, the leakage becomes negligible, and the structure supports a dominant TE_{10} mode similar to that in a rectangular waveguide.

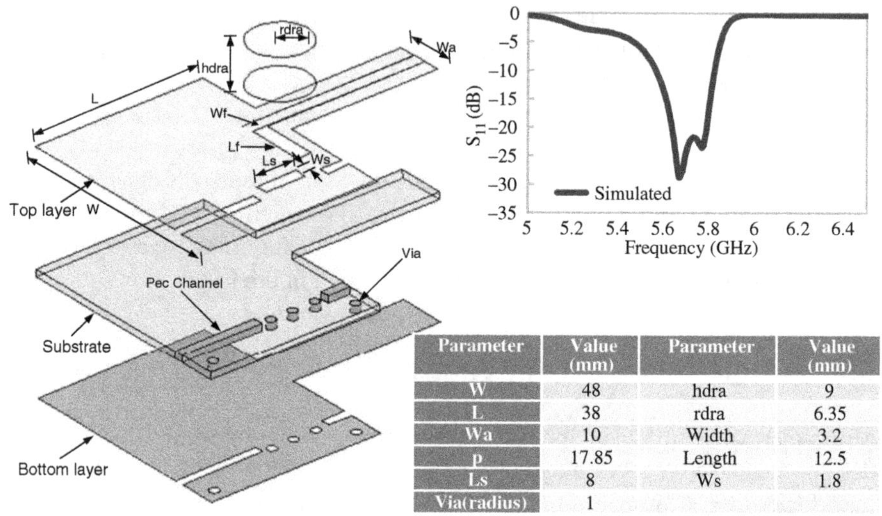

Parameter	Value (mm)	Parameter	Value (mm)
W	48	hdra	9
L	38	rdra	6.35
Wa	10	Width	3.2
p	17.85	Length	12.5
Ls	8	Ws	1.8
Via(radius)	1		

Figure 5.2 SIW-DRA-based single-band antenna. *Source*: Naeem et al. (2017)/John Wiley & Sons/CC BY 4.0.

The resonant frequency of a single-band SIW cavity is determined mainly by W_{eff}. For the TE_{10} mode, the cutoff frequency is:

$$f_c = \frac{c}{2W_{\text{eff}}\sqrt{\varepsilon_r}}, \tag{5.2}$$

where c is the speed of light in vacuum, and ε_r is the relative permittivity of the substrate. For a radiation element designed to operate at f_0, the cavity width is selected so that f_0 is above cutoff, but still within a range where the fields remain well confined.

5.2.2 Rules for via Placement

In SIW components, the via rows are critical. They define the boundaries of the waveguide and determine how much unwanted radiation leaks through the gaps. Practical rules have been established based on full-wave results and measurements. A typical design follows three conditions:

1. The via pitch must satisfy:

$$p \leq 2d, \tag{5.3}$$

which keeps the gap between vias small enough to limit leakage.

2. The via pitch must also satisfy:

$$p \leq \frac{\lambda_g}{5}, \tag{5.4}$$

where λ_g is the guided wavelength inside the SIW at the operating frequency. This ensures that the periodic via fence approximates a continuous wall from an electromagnetic point of view.

3. The via diameter is usually chosen within $0.1W_{\mathrm{SIW}}$ to $0.15W_{\mathrm{SIW}}$. Smaller vias tend to increase loss, while very large vias reduce the available area for the waveguide interior.

These rules allow the SIW to behave almost identically to a metallic waveguide, while still being fabricated using a standard PCB process. For mm-wave designs, via tolerance and plating quality become important, but modern fabrication can meet these requirements reliably.

5.2.3 Cavity-type Single-band SIW Antenna

A common single-band SIW antenna is a cavity-backed slot. The cavity resonates at one frequency, and the slot radiates when placed at a point where the magnetic field is strong. To approximate this resonance, the cavity length L is selected so that it fits half a guided wavelength:

$$L \approx \frac{\lambda_g}{2} = \frac{c}{2f_0\sqrt{\varepsilon_{\mathrm{eff}}}}, \tag{5.5}$$

where $\varepsilon_{\mathrm{eff}}$ is the effective permittivity accounting for the field distribution in the SIW. Although the expression resembles that of microstrip resonators, the SIW version is more stable because the field is largely enclosed.

The slot is normally cut in the top metal layer. It couples to the TE_{10} field inside the cavity, which has a maximum magnetic field along the sidewalls and an electric field concentrated between them. A longitudinal slot placed slightly offset from the centerline provides efficient radiation without disturbing the dominant mode. The slot length is typically close to half a free-space wavelength:

$$L_s \approx \frac{\lambda_0}{2}, \tag{5.6}$$

although adjustments are made to account for the cavity loading and fringing-field effects.

At mm-wave frequencies, several feed options exist for a single-band SIW antenna. A common approach is a microstrip-to-SIW transition. It gradually transforms the quasi-TEM mode of a microstrip line into the TE_{10} mode of the SIW. The taper is shaped to avoid abrupt impedance steps. As long as the transition length is around a quarter of the guided wavelength or more, reflections remain small.

Another option is an inset coaxial probe. The probe excites the cavity directly, but it introduces parasitics that become difficult to control at mm-wave frequencies. For high-volume PCB-based antennas, planar transition tends to be the preferred choice.

5.2.4 Radiation and Bandwidth Considerations

Single-band SIW antennas usually operate over a narrow bandwidth because the cavity behavior is dominated by a single resonant mode. The bandwidth depends on the cavity Q-factor, which is influenced by dielectric losses, conductor losses, and radiation through the slot. A higher dielectric constant reduces size but narrows the bandwidth. A lower-loss substrate improves efficiency but increases cost.

Radiation efficiency benefits from the enclosed fields inside the SIW layer. The enclosure suppresses surface waves that often degrade patterns in microstrip-fed antennas. The result is a more stable gain and a predictable main beam. For mm-wave radar and point-to-point links, this consistency is often more important than wide bandwidth.

5.2.5 Summary of Design Process

A single-band SIW antenna can be designed through the following steps:

1. Select operating frequency f_0.
2. Choose substrate thickness and permittivity.
3. Determine W_{eff} from the cutoff condition.
4. Apply the correction term to obtain the physical width W_{SIW}.
5. Choose via diameter and pitch to satisfy the leakage rules.
6. Set the cavity length to approximately half the guided wavelength.
7. Place the slot to achieve the desired radiation level.
8. Evaluate the antenna matching and radiation performance.
9. Fabricate the antenna geometry. Various methods can be used for fabrication.
10. Measure the antenna matching and radiation performance. Compare the simulated and measured data.

5.3 Multiband SIW Antennas

Millimetre-wave systems often need to operate at more than one frequency band. Radar modules, short-range communication links, and integrated sensing platforms increasingly combine bands to support different functions on the same hardware. SIW technology is well-suited for this because it can host multiple resonant paths inside a single structure without relying on external filters or additional feed lines. Multiband SIW antennas draw on the cavity behavior of the waveguide, allowing separate modes or separate resonant sections to radiate independently while sharing the same physical footprint (Ashraf et al., 2015).

5.3.1 Concept of Multiband Operation

A single SIW cavity supports a fundamental TE_{10} mode, but higher-order modes can also be used to create additional resonances. By shaping the cavity, introducing internal partitions, or modifying the slot arrangement on the top conductor, several resonant frequencies can be obtained. Each band can radiate from either a dedicated slot or a shared aperture, depending on how the internal fields couple to the radiating structure.

In multiband antennas, the key challenge is achieving separation between operating frequencies without relying on bulky external diplexers. SIW makes this possible because the geometry of the cavity controls the modal distribution predictably. When two or more modes are sufficiently separated in frequency and have distinct field profiles, each can be tapped selectively by placing slots at points of strong magnetic or electric field for the targeted mode.

5.3.2 Self-diplexing SIW Antennas

A self-diplexing SIW antenna radiates at two well-separated frequency bands using a single physical aperture (Nandi and Mohan, 2017; Pradhan et al., 2023). Instead of external diplexers, the antenna relies on two mechanisms:

1. Distinct cavity resonances produced by structural modifications, and
2. Spatial location of slots that selectively excite one mode over another.

One common approach is to divide the cavity into two unequal regions using a short internal wall or a row of vias. Each region behaves as a smaller SIW resonator with its own dominant-mode frequency:

$$f_{c,i} = \frac{c}{2W_{\text{eff},i}\sqrt{\varepsilon_r}}, \quad i = 1,\ 2, \tag{5.7}$$

where $W_{\text{eff},i}$ is the effective width of the ith resonant section. The two sections couple through a narrow opening or a short slot, enabling both bands to share a single radiating aperture. The designer adjusts the widths and lengths until the two desired resonances are obtained.

Another diplexing method places two slots on the top metal. Each slot is positioned to align with the magnetic-field maxima of the corresponding mode, so each band radiates primarily through its intended slot. When done correctly, unwanted

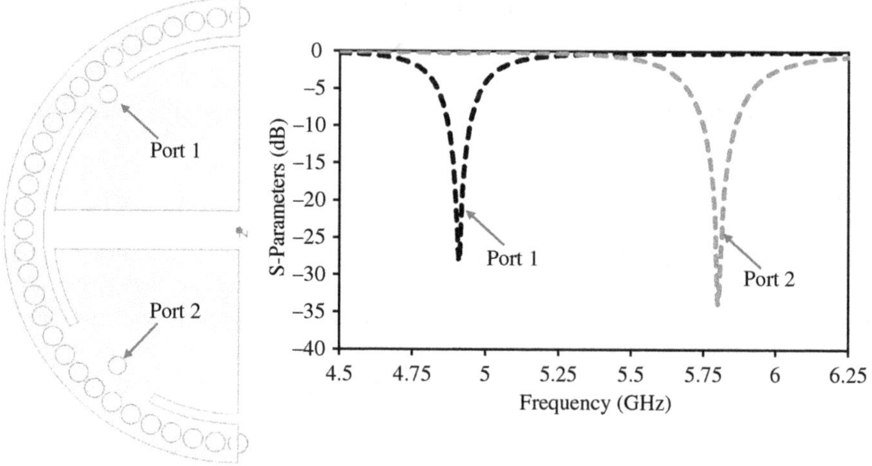

Figure 5.3 SIW-based self-diplexing dual-band antenna.

cross-coupling remains low without the need for distributed filter sections. An example of a self-diplexing antenna using QMSIW cavities is shown in Figure 5.3 (Mallat and Iqbal, 2022). It can be seen that the antenna has two QMSIW resonators. Both QMSIW resonators are fed using coaxial feeding. The reflection coefficient of the antenna is shown in the same figure.

5.3.3 Self-triplexing SIW Antennas

Extending the approach to three frequency bands is possible by using three resonant paths or by exploiting two higher-order modes in addition to the fundamental mode (Nigam et al., 2021; Priya and Dwari, 2020). A triplexing SIW antenna typically employs one of two layouts:

- A single cavity with three distinct resonant modes, or
- A merged structure with three sub-cavities connected by narrow SIW channels.

When the design relies on modal behavior, the cavity is shaped so that the TE_{10}, TE_{20}, and a perturbed higher-order mode appear at frequencies that match the desired operating bands. The slot arrangement then becomes critical. Each slot is placed so that it couples strongly to one mode but weakly to the others. This requires careful examination of field distributions obtained from full-wave simulation. Even small changes in slot offset or length can shift the radiating efficiency of a band, so the tuning step is often iterative.

The multi-cavity method is more flexible. Each sub-cavity is dimensioned to produce a single resonance:

$$L_i \approx \frac{\lambda_{g,i}}{2}, \quad i = 1, 2, 3, \tag{5.8}$$

with $\lambda_{g,i}$ calculated from the SIW waveguide dispersion relation for the chosen frequency. The sub-cavities are linked to a shared aperture. If the interconnecting windows are sufficiently narrow, the energy at each band remains mostly confined to its intended region. This arrangement simplifies tuning because each section behaves like an independent resonator. An example of a self-triplexing antenna using SIW cavities is shown in Figure 5.4 (Iqbal et al., 2020). It can be seen that the antenna has three slot resonators in the SIW cavity. The reflection coefficient of the antenna is shown in the same figure, showing three resonances.

5.3.4 Self-quadruplexing SIW Antennas

Four-band SIW antennas push the structure further. At this point, a single cavity usually becomes difficult to control, so designers rely on either:

1. Multiple sub-cavities arranged side by side, or
2. Perturbation techniques that produce four resonances within a unified SIW platform.

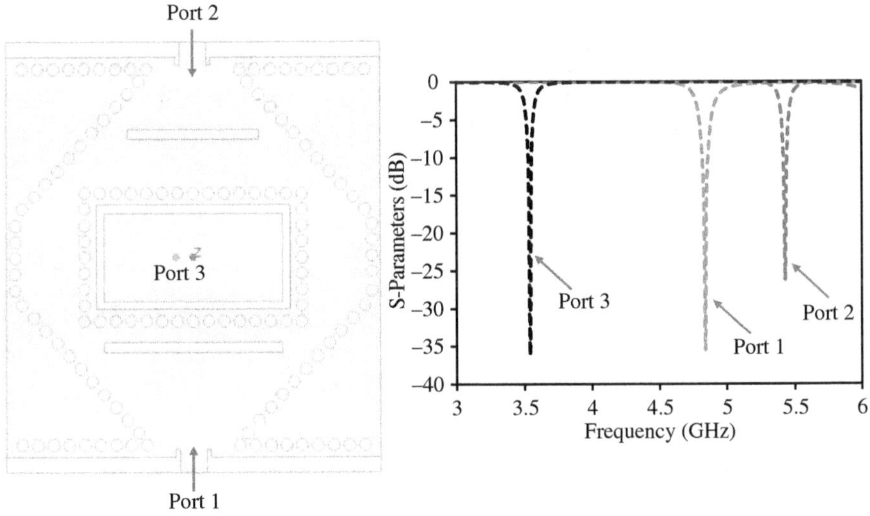

Figure 5.4 SIW-based self-triplexing triple-band antenna.

For the multi-cavity form, each region is sized for a different frequency, and all are connected to a common radiating slot or aperture. The tuning process focuses on controlling the small coupling apertures so that energy at each band remains isolated. The approach scales reasonably well when the target frequencies are widely spaced (Barik and Koziel, 2023; Kumar, 2019).

The perturbed-cavity form uses slot loading, stepped cavity widths, or embedded posts to generate additional modes. The positions of these features shift the modal frequencies according to how they alter the field confinement. Although more compact, these designs need thorough simulation because higher-order modes may interact in unexpected ways if the perturbations are not placed carefully. An example of a self-quadruplexing antenna using SIW cavities is shown in Figure 5.5 (Naseri et al., 2022). It can be seen that the antenna has four resonators in the SIW cavity. The reflection coefficient of the antenna is shown in the same figure, showing four resonances.

5.3.5 Placement of Vias and Multiband Stability

The role of the via rows becomes even more important in multiband designs. The basic rules for leakage still apply:

$$p \leq 2d, \qquad p \leq \frac{\lambda_g}{5}, \tag{5.9}$$

but multiband cavities are more sensitive to unwanted losses because each additional resonance introduces another point where energy can leak or couple into parasitic modes. Maintaining consistent via spacing around the entire cavity helps stabilize all resonant frequencies. Any local irregularity in the via fence can shift one of the bands or distort its radiation pattern.

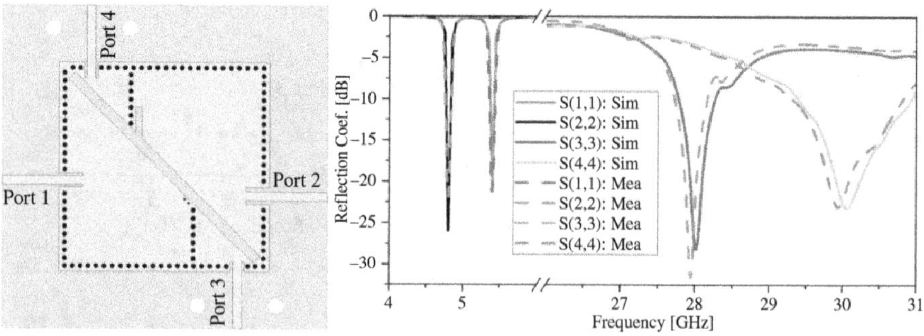

Figure 5.5 SIW-based microwave and mm-wave self-quadruplexing antenna configuration.

When sub-cavities are used, vias define the internal boundaries. Their alignment and diameter must be controlled so that the electrical length of each section matches its intended design. Even small deviations can introduce unexpected coupling between bands.

5.3.6 Feed Structures and Isolation Between Bands

Multiband SIW antennas usually rely on a single input port, so isolation between bands depends heavily on the natural filtering behavior of the resonators. In diplexing and triplexing designs, the feed is placed at a point where the input coupling is strong for all modes. The separation in frequency ensures that each band sees a different effective impedance at the feed point.

For quadruplexing antennas, the feed location and geometry must be chosen so that it couples to each intended mode without favoring an unintended one. Short tapered transitions or compact SIW transformers are sometimes added to balance the input match across all four frequencies.

5.3.7 Advantages and Application Context

Self-diplexing, triplexing, and quadruplexing SIW antennas are attractive when space is limited and cost must be controlled. Automotive radar modules, short-range sensing nodes, and compact communication terminals benefit from being able to host multiple bands within a single antenna block. The reduced number of external components simplifies assembly and improves reliability, especially in systems exposed to vibration or temperature swings.

The strong field confinement inside the SIW layers reduces surface-wave coupling between bands, contributing to stable isolation.

5.4 SIW Antennas Arrays

In the majority of applications, a high gain and directed beam are required at mm-wave frequencies. These attributes are traditionally achieved by using antenna arrays. SIW antennas are a good choice for the mm-wave RF circuitry chain. SIW possesses many advantages of a classical metallic waveguide in planar form. With current advancements, integration of components is vital. SIW circuits are planar and can be integrated with other RF circuits and components. Achieving a high-gain and directed beam is difficult using a single radiator. This can be achieved by arranging radiators in a specific order to form an antenna array. Antenna arrays have several integrated radiators, which provide controlled beam, directed pattern, high gain, and improved link margins (Klionovski et al., 2023). An SIW-based array configuration is illustrated in Figure 5.6.

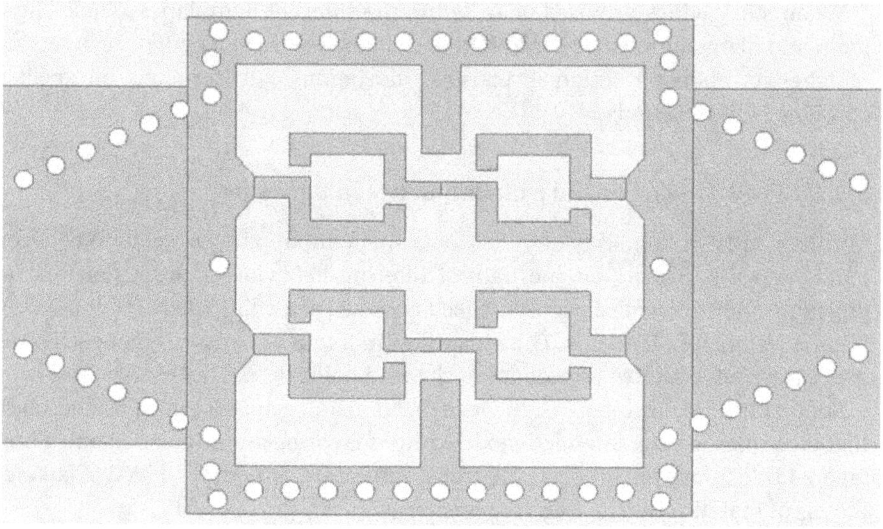

Figure 5.6 SIW-based mm-wave antenna array configuration.

In most applications, high gain and a directed beam are required at mm-wave frequencies. These attributes are traditionally achieved by using antenna arrays. SIW antennas are a good choice for the mm-wave RF circuitry chain. SIW possesses many advantages of a classical metallic waveguide in planar form. With current advancements, integration of components is vital. SIW circuits are planar and can be integrated with other RF circuits and components. Achieving high gain and a directed beam is difficult using a single radiator. This can be achieved by arranging radiators in a specific order to form an antenna array. Antenna arrays have several integrated radiators that provide a controlled beam, a directed pattern, high gain, and improved link margins.

5.4.1 Array Configurations

Most SIW arrays rely on either a corporate-feed architecture or a series-feed structure. Corporate feeding uses a branching power-division network built within the SIW layer. This method provides equal-phase excitation to each element and offers strong control over amplitude distribution. It is widely used when the goal is to achieve symmetric main beams and low sidelobe levels. The cost is a larger footprint and increased cumulative loss along the feed lines.

Series feeding distributes energy sequentially from one element to the next one. This topology is useful for generating simpler layouts and small circuit dimensions. These attributes are useful for mm-wave RF circuits. Despite several advantages, it has some limitations. For example, controlling the amplitude and phase of each array element is challenging. Controlling phase and amplitude is an important aspect of beam steering, which is crucial for mm-wave RF circuits. Despite these limitations, it is still considered an attractive choice for compact mm-wave RF devices.

Slot arrays are common in SIW structures. A line of radiating slots etched on the top metal couples energy from the guided mode inside the SIW to free space. The element spacing usually follows approximately half the guided wavelength, which is shorter than the free-space wavelength. This allows a larger number of elements within a given aperture size, increasing gain without extending antenna dimensions (Klionovski et al., 2023).

Patch-based SIW arrays are also widely used. Each SIW cavity feeds a patch through a coupling slot or probe. The cavity dimensions and coupling strength help tune the element bandwidth and improve impedance stability. These arrays offer good control over the radiation response and can support broadband operation when designed carefully.

5.4.2 Beam Steering and Pattern Control

One advantage of SIW arrays is the ability to steer the beam electronically or through structural phase adjustments. By applying controlled phase shifts across the elements, the main beam can be directed toward different angles. At mm-wave frequencies, small phase differences produce notable steering because of the short wavelength. This allows precise control over scan ranges without needing large physical adjustments.

Phase shifts can be implemented using delay lines inside the SIW, via-loaded sections, or tunable dielectric layers. Amplitude tapering is often used to control sidelobe levels. Because SIW networks confine fields effectively, the amplitude delivered to each element is consistent, which helps maintain accurate taper profiles. Standard tapers such as Taylor or Chebyshev distributions are often applied to improve pattern performance.

5.4.3 Mutual Coupling Considerations

Although SIW structures confine the fields more strongly than microstrip lines, mutual coupling still affects array behavior, especially when the spacing between elements becomes small. High coupling can distort the impedance response and alter the radiation pattern. Designers often control coupling through slot offsets, cavity spacing, or by inserting additional vias between array elements to suppress unwanted modes.

Parasitic structures, such as slots or patches placed between active elements, can redistribute the fields and help reduce coupling. At mm-wave frequencies, these small structural adjustments can have a substantial influence on the final radiation characteristics. The shielding action of SIW walls further aids in maintaining consistent inter-element isolation.

5.4.4 Performance at mm-Wave Frequencies

At mm-wave frequencies, conductor and dielectric losses become more pronounced. SIW structures help mitigate these losses by confining energy inside the dielectric and limiting radiation leakage. The waveguide-like behavior provides a stable phase response across the feed network, improving the accuracy of the array's beam formation.

SIW arrays can also be fabricated on cost-effective materials without requiring high-end substrates. While performance improves with low-loss dielectrics, many practical SIW arrays built on mid-range substrates still achieve acceptable efficiency and gain. Their balance of cost, performance, and integration capability explains their continued adoption in modern mm-wave systems.

5.5 Wideband SIW Antennas

Wideband antennas based on substrate-integrated waveguide (SIW) structures have become increasingly important as modern wireless systems move toward higher data rates and more flexible spectrum use. In many mm-wave front ends, narrowband radiators no longer meet system requirements. Communication links must tolerate frequency drift, fabrication variations, and multichannel operation, while radar and sensing platforms often sweep across several gigahertz of bandwidth. These demands have pushed designers to explore SIW geometries that can support wider operating ranges without losing the advantages of waveguide-like confinement (Aliakbari et al., 2019; Araghi et al., 2022; Zhang et al., 2020).

A key challenge comes from the modal nature of SIW cavities. Traditional SIW structures support well-defined waveguide modes that tend to be selective in frequency. To broaden the usable band, designers modify the field distribution or introduce multiple resonant paths within the antenna. Techniques such as tapered slots, stepped transitions, modified apertures, and multi-resonant cavities are commonly employed. Because SIW structures are implemented with standard printed-circuit processes, these features can be integrated without adding mechanical complexity.

One widely used method for achieving wideband behavior is the tapered slot radiator fed by an SIW line. The gradual taper transforms the guided mode into a radiated wave with minimal discontinuity, which reduces reflections and smooths the impedance profile over a large frequency span. Such designs often provide the bandwidth required for mm-wave radar and multiband communication systems. Another approach uses ridged or stepped SIW channels. By adjusting the height or introducing internal loading, additional inductive or capacitive effects appear, creating multiple resonances that merge into a broader band.

Coupling-based techniques are also effective. Some wideband SIW antennas rely on arrays of slots with different lengths or orientations. Each slot supports a slightly different resonant response, and their combined effect extends the operating band. When placed carefully, these slots preserve stable radiation characteristics while keeping the antenna compact. Patch-like apertures coupled to SIW cavities offer another option. By tuning the coupling level and cavity geometry, overlapping resonances can be produced, resulting in a continuous wideband response.

Maintaining stable radiation patterns across the operating band is another important consideration. Wideband designs sometimes suffer from pattern distortion at the band edges. To counter this, symmetrical layouts or distributed radiating apertures are used to balance the field distribution. The SIW walls help suppress surface-wave propagation, which improves pattern consistency and reduces unwanted coupling to adjacent circuit blocks.

5.6 Conclusion

This chapter covered SIW technology in detail and explained its importance for mm-wave frequencies. In fact, SIW contains the advantages of traditional metallic waveguides in a planar form. This chapter has explained why SIW technology is better compared to other planar technologies. SIW is a better choice despite the antenna having single resonances, multiple resonances, or wideband performance. Also, several advanced and timely topics of SIW-based self-diplexing, self-triplexing, and self-quadruplexing antennas have been discussed. Overall, SIW provides numerous advantages and is well-suited for mm-wave systems.

References

Hanieh Aliakbari, Mohammad Mosalanejad, Charlotte Soens, Guy A.E. Vandenbosch, and Buon Kiong Lau. Wideband SIW-based low-cost multilayer slot antenna array for *e*-band applications. *IEEE Transactions on Components, Packaging and Manufacturing Technology*, 9(8):1568–1575, 2019. https://doi.org/10.1109/TCPMT.2019.2910385.

Mohammad Alibakhshikenari, Bal S Virdee, Shahram Salekzamankhani, Sonia Aïssa, Chan H See, Navneet Soin, Sam J Fishlock, Ayman A Althuwayb, Raed Abd-Alhameed, Isabelle Huynen, et al. High-isolation antenna array using SIW and realized with a graphene layer for sub-terahertz wireless applications. *Scientific Reports*, 11(1):10218, 2021.

Ayman A Althuwayb. MTM-and SIW-inspired bowtie antenna loaded with AMC for 5G mm-wave applications. *International Journal of Antennas and Propagation*, 2021(1):6658819, 2021

Ali Araghi, Mohsen Khalily, Pei Xiao, Rahim Tafazolli, and David R. Jackson. Long slot mmwave low-SLL periodic-modulated leaky-wave antenna based on empty SIW. *IEEE Transactions on Antennas and Propagation*, 70(3):1857–1868, 2022. https://doi.org/10.1109/TAP.2021.3137183.

Nadeem Ashraf, Osama Haraz, Muhammad A. Ashraf, and Saleh Alshebeili. 28/38-GHz dual-band millimeter wave SIW array antenna with EBG structures for 5G applications. In *2015 International Conference on Information and Communication Technology Research (ICTRC)*, pages 5–8, 2015. https://doi.org/10.1109/ICTRC.2015.7156407.

Rusan Kumar Barik and Slawomir Koziel. Highly miniaturized self-quadruplexing antenna based on substrate-integrated rectangular cavity. *IEEE Antennas and Wireless Propagation Letters*, 22(3):482–486, 2023. https://doi.org/10.1109/LAWP.2022.3216305.

Yaling Chen, Long Zhang, Yejun He, Chunxu Mao, Sai-Wai Wong, Wenting Li, Peng Chu, and Steven Gao. Broadband high-gain SIW horn antenna loaded with tapered multistrip transition and dielectric slab for mm-wave application. *IEEE Transactions on Antennas and Propagation*, 70(7):5947–5952, 2022. https://doi.org/10.1109/TAP.2022.3161349.

Amjad Iqbal, Marwa Ammar Selmi, Lway Faisal Abdulrazak, Omar A. Saraereh, Nazih Khaddaj Mallat, and Amor Smida. A compact substrate integrated waveguide cavity-backed self-triplexing antenna. *IEEE Transactions on Circuits and Systems II: Express Briefs*, 67(11):2362–2366, 2020. https://doi.org/10.1109/TCSII.2020.2966527.

Kirill Klionovski, Hanguang Liao, Sergey E. Bankov, Zubair Akhter, and Atif Shamim. A planar SIW-based mm-wave frequency-scanning slot antenna array with no scan blindness at normal. *IEEE Transactions on Antennas and Propagation*, 71(1):563–569, 2023. https://doi.org/10.1109/TAP.2022.3215856.

Arvind Kumar. Design of self-quadruplexing antenna using substrate-integrated waveguide technique. *Microwave and Optical Technology Letters*, 61(12):2687–2689, 2019.

Priyanka Kumawat and Sunil Joshi. Review of slotted SIW antenna at 28 GHz and 38 GHz for mm-wave applications. In *2020 12th International Conference on*

Computational Intelligence and Communication Networks (CICN), pages 8–13, 2020. https://doi.org/10.1109/CICN49253.2020.9242587.

Xiangxiang Li, Yeqiang Li, Zhe Li, Qinglong Kong, Hailin Wang, and Guoqiang Lv. Wideband high-gain SIW-fed antenna array for mm-wave applications. *Microwave and Optical Technology Letters*, 62(3):1341–1351, 2020.

Nazih Khaddaj Mallat and Amjad Iqbal. Substrate integrated waveguide-based simultaneous transmit and receive antenna for full-duplex wearable devices. *International Journal of RF and Microwave Computer-Aided Engineering*, 32(7):e23188, 2022.

Umair Naeem, Amjad Iqbal, Muhammad Farhan Shafique, and Stéphane Bila. Efficient design methodology for a complex DRA-SIW filter-antenna subsystem. *International Journal of Antennas and Propagation*, 2017(1):6401810, 2017.

Sourav Nandi and Akhilesh Mohan. An SIW cavity-backed self-diplexing antenna. *IEEE Antennas and Wireless Propagation Letters*, 16:2708–2711, 2017. https://doi.org/10.1109/LAWP.2017.2743017.

Hassan Naseri, Peyman PourMohammadi, Amjad Iqbal, Ahmed A. Kishk, and Tayeb A. Denidni. Siw-based self-quadruplexing antenna for microwave and mm-wave frequencies. *IEEE Antennas and Wireless Propagation Letters*, 21(7):1482–1486, 2022. https://doi.org/10.1109/LAWP.2022.3172007.

Padmini Nigam, Arjuna Muduli, Sandeep Sharma, and Amrindra Pal. Tri-fed SIW based self-triplexer antenna for X-band applications. *International Journal of Applied Electromagnetics and Mechanics*, 67(1):97–112, 2021.

Amit Kr. Pandey, Nikesh Kumar Sahu, Ravi Kumar Gangwar, and Raghvendra Kumar Chaudhary. SIW-cavity-backed wideband circularly polarized antenna using modified split-ring slot as a radiator for mm-wave IoT applications. *IEEE Internet of Things Journal*, 11(7):11793–11799, 2024. https://doi.org/10.1109/JIOT.2023.3332169.

Peyman PourMohammadi, Hassan Naseri, Noureddine Melouki, Fahad Ahmed, Amjad Iqbal, Guy A. E. Vandenbosch, and Tayeb A. Denidni. Compact SIW-based self-quadruplexing antenna for microwave and mm-wave communications. *IEEE Transactions on Circuits and Systems II: Express Briefs*, 70(9):3368–3372, 2023. https://doi.org/10.1109/TCSII.2023.3264708.

Nrusingha Charan Pradhan, Mettu Goutham Reddy, Karthikeyan Sholampettai Subramanian, Rusan Kumar Barik, Slawomir Koziel, and Qingsha S. Cheng. Microfluidic SIW-based tunable self-diplexing antenna for sub-6 GHz band applications. *IEEE Transactions on Circuits and Systems II: Express Briefs*, 70(4):1435–1439, 2023. https://doi.org/10.1109/TCSII.2022.3224234.

Shruti Priya and Santanu Dwari. A compact self-triplexing antenna using HMSIW cavity. *IEEE Antennas and Wireless Propagation Letters*, 19(5):861–865, 2020. https://doi.org/10.1109/LAWP.2020.2982533.

Qiang Sun, Yong-Ling Ban, Ji-Wei Lian, Yanhui Liu, and Zaiping Nie. Millimeter-wave multibeam antenna based on folded c-type SIW. *IEEE Transactions on Antennas and Propagation*, 68(5):3465–3476, 2020. https://doi.org/10.1109/TAP.2020.2966050.

Long Zhang, Ke Wu, Sai-Wai Wong, Yejun He, Peng Chu, Wenting Li, Kai Xu Wang, and Steven Gao. Wideband high-efficiency circularly polarized SIW-fed S-dipole array for millimeter-wave applications. *IEEE Transactions on Antennas and Propagation*, 68(3):2422–2427, 2020. https://doi.org/10.1109/TAP.2019.2940468.

6

Wideband Millimeter-wave Antennas

Wide bandwidth is a key need in mm-wave systems. It is not just a design choice. It is required for the goals of future wireless systems. In this chapter, we discuss wideband antennas and explain their importance in millimeter-wave (mm-wave) communication systems. We then present the most widely used bandwidth enhancement techniques.

6.1 Why Wideband Antennas?

Wide bandwidth is very important in mm-wave antennas. It gives the antenna the power to carry more information. When the bandwidth is small, the antenna can only send a small amount of data. This was acceptable many years ago when people only sent voice or simple messages. Today, however, the world is different. Nowadays, users routinely stream high-definition content, including 4K and 8K video. They use virtual reality and play online games. They upload and download large files. All these needs ask for a fast speed. Fast speed is only possible when the antenna supports a wide bandwidth. More bandwidth means more data can move at the same time. It feels like having a wider road where more cars can travel without traffic (Chen, 2013; Cicchetti et al., 2017; Yazdandoost and Kohno, 2004).

Another reason is that wireless systems are now very crowded. Several wireless nodes share the same frequency band simultaneously. For example, all nodes, such as phones, computers, cameras, and many others, want to have seamless communication. If the transmitting or receiving antenna has limited bandwidth, the network may become inefficient, disconnect frequently, and be slow. In contrast, antennas with a broad bandwidth provide several chunks of channels. With a number of small channels, users can easily connect to the network and use it efficiently without waiting for long. It provides an opportunity for all users to stay connected to the same network without frequent disconnections (Balani et al., 2019; Saeidi et al., 2019).

The signals operating at high frequencies (mm-wave) have limited area coverage as they dissipate quickly in the air. Moreover, they are unable to effectively transmit through obstacles or the human body. In these cases, a major portion of power is reflected or refracted and does not reach its target. Therefore, signals should be strong and resilient enough to operate in the mm-wave frequency range. Antennas and networks having a broad bandwidth are widely used to tackle these problems by using advanced signal processing. They can also be used for radiation pattern reconfigurability by directing the beam in the desired direction. The wireless link is greatly impacted by the bandwidth. It has a direct relationship with the wireless link between the network and the user. In fact, wideband provides a more reliable and stable communication link (Shen and Win, 2010).

Many new technologies also require low latency to ensure high performance. Industrial robots, remote surgery, and self-driving cars all must act quickly. They can't afford the time taken up by slow communication. A large bandwidth helps to reduce the latency. Data transfers fast between devices, and there will not be increased delay due to limited bandwidth, which might induce errors. A wide bandwidth offers greater confidence in real-time systems in terms of performance and safety.

Millimeter-wave systems also use new features like large multiple-input multiple-output (MIMO), fast beam steering, and even sensing the environment at the same time as communication. These features need a lot of frequency bands. They cannot work properly if the antenna is narrowband. A narrow-bandwidth antenna limits many advanced functions. Apart from a lower data rate, it also introduces a problem at the system-level hardware. An introduction of a new service may require new hardware, and the existing one should be discarded.

Antennas having broad bandwidths are vital for future technologies. It is well-known that the number of users increases each year drastically. Similar to the increase in users, the demand for high speed and more data also increases. When systems have the ability to operate over a wide range of frequencies, several users can be accommodated simultaneously. Moreover, with a wider operational bandwidth, the existing hardware can be used for new services without altering or introducing alternative hardware. The same hardware can be used for several years without modifying it. It reduces the operational costs for new services. In the case of narrowband, the outdated hardware should be replaced with new hardware. Technologies operating at mm-wave frequencies change rapidly; therefore, a wideband operational bandwidth is crucial.

To summarize the above thoughts, wideband is vital for high-speed communication compared to narrowband. Narrowband communication systems cannot support high data rates. Moreover, it has restrictions on connected users. Narrowband communication leads to short-range communication and a weak link margin. It has limitations in adopting next-generation technologies such as beamforming.

Based on these comments, narrowband is not considered suitable for high-speed communication in the mm-wave frequency band. The main reason for using the mm-wave frequency range is to provide high-speed communication with fewer errors and high performance. As a solution to this issue, wideband antennas can be employed to overcome the limitations of narrowband antennas (Kramer et al., 2009; Kumar and Kumar, 2019; Lee et al., 2005).

6.2 Bandwidth Enhancement Techniques

Bandwidth enhancement techniques improve the performance of mm-wave antennas. Designers use these methods because the natural bandwidth of a small antenna is usually not enough. One simple way is to change the antenna shape. When the shape has extra parts like slots, stubs, or tapered edges, the antenna can create more than one resonance. The excitation of multiple modes is necessary for generating several resonances. In fact, these modes are coupled together to obtain a wider bandwidth. Proper placement of slots, stubs, or tapered edges is vital for efficient coupling. A very similar approach for obtaining a wide operational bandwidth is by coupling a parasitic radiator with the main antenna. If both the parasitic structure and antenna element are planar, the gap between them defines the coupling coefficient. In such topologies, a few modes are excited within the main radiator and coupled with the excited modes in the parasitic radiator. It is worth mentioning that most of the parasitic elements are capacitively coupled with no direct energy being injected into them (Akinola et al., 2019; Kumar et al., 2016; Mishra et al., 2022).

The transmission lines are important for antenna matching. The transmission lines are normally designed at 50Ω and connected to the antenna radiator. A proper match between the feed and the antenna radiator improves impedance matching and reduces reflections. Transmission lines with taper and wide feed are found to be better in terms of wideband frequency matching. Another widely used feeding structure for wideband mm-wave antennas is the coplanar waveguide (CPW). This type of topology handles high frequencies with less leakage. Antenna and RF engineers also suggest using low-loss material for mm-wave frequencies. Low-loss material helps in obtaining a wider operational bandwidth. Using low-loss substrates and better metal surfaces can protect the signal and keep the bandwidth wide. Air gaps are sometimes added to reduce dielectric loss, but they must be controlled well.

Suitable mutual coupling between antenna arrays also increases operational bandwidth. When many small antennas work together, the overall response becomes smoother and wider. But spacing and size must be properly designed to achieve the desired performance. If not, the antennas may disturb each other and reduce performance. Vias and ground structures are also used to control surface waves. This helps more of the energy radiate in the right direction, so the antenna works well across the band.

In the end, no single method is enough on its own. Designers combine many small improvements. They change the geometry, improve the feed, choose better materials, and control the ground and surface waves. With careful design, the bandwidth can be made sufficiently wide for modern mm-wave systems (Akinola et al., 2019; Kumar et al., 2016; Mishra et al., 2022; Nahar and Rawat, 2022).

6.3 Stacked Microstrip Antenna Techniques for Bandwidth Enhancement

The stacked microstrip technique is a widely used method for enhancing bandwidth in mm-wave systems. A normal microstrip patch antenna typically exhibits a narrow bandwidth. It generally operates based on a single dominant resonant mode. This is inadequate for modern systems that need a wide frequency range. The stacked design improves this problem by adding one or more extra patches above the main patch. These extra patches create new resonances. When these resonances join with the main one, the final bandwidth becomes wider. The design procedure is not particularly complex and can be easily implemented. The size also stays compact. For this reason, stacked antennas are widely used in many mm-wave devices today (AbuTarboush et al., 2009; Mohammad et al., 2006; Nishiyama et al., 2004). An example of stacked microstrip patch antennas is shown in Figure 6.1.

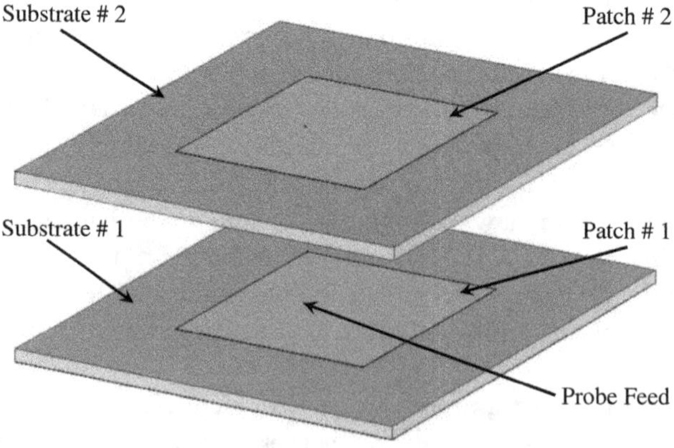

Figure 6.1 An example of a two-layer microstrip patch antenna excited using the probe feeding technique.

6.3.1 Basic Structure and Bandwidth Problem

A microstrip patch has a metal layer on a dielectric substrate. It has a ground plane under the substrate. It acts like a resonant cavity. The bandwidth depends on the thickness of the substrate and on the dielectric constant. A thin substrate and a high dielectric constant reduce the bandwidth. In fact, antennas having high permittivity or high Q-factor store most of the energy and radiate less in the air. For enhanced bandwidth, the stored energy should be less in amount than the radiated energy.

6.3.2 Principle of Bandwidth Enhancement

The excited current and its flow widely change when the stack configuration is adopted. Both stacks radiate energy simultaneously, which increases the radiated power compared to a standalone configuration. More specifically, the radiated power is more than the stored one. In a normal configuration, one element is directly fed through a transmission line or coax feed. The second resonator is coupled capacitively, inductively, or both, which becomes strong at the new resonant frequencies. When the designer adjusts the size and height of the top patch, the resonances overlap. The antenna stays matched to the feed across a wider band.

The bandwidth BW of an antenna is related to the quality factor Q. A simple relation is:

$$\mathrm{BW} \approx \frac{1}{Q}.$$

A lower value of Q gives a wider bandwidth. A stacked antenna has a lower Q because it radiates more energy.

6.3.3 Coupling Distance and Material Effects

The gap between patches is important. If the patches are too close, the coupling becomes too strong, and the antenna becomes unstable. If the patches are too far, the coupling becomes weak, which may reduce bandwidth enhancement effects. Many designers use a distance close to one-tenth of the wavelength inside the substrate. The dielectric constant also affects the design. A low dielectric constant increases bandwidth but also increases the size. A high dielectric constant makes the antenna small but reduces bandwidth. A stacked design helps balance these effects. The lower patch can use a high-dielectric material to stay small. The upper patch can use a low-dielectric material to help with bandwidth.

The wavelength inside the material relates to the free space wavelength λ_0:

$$\lambda = \frac{\lambda_0}{\sqrt{\varepsilon_r}}.$$

A lower dielectric constant ε_r increases radiation and helps bandwidth.

6.3.4 Feeding Techniques

The choice and design of the feeding structure play a critical role in determining the achievable impedance bandwidth. A probe feed may create extra inductance. A microstrip line feed is easy to make, but it may give weak coupling to the top layer. Aperture coupling is very useful in stacked antennas. A slot in the ground plane sends energy from the feed line into the lower patch. The lower patch then couples to the upper patch. This creates multiple resonances in a simple way, which helps broaden the overall impedance bandwidth.

6.3.5 Mechanical and Manufacturing Considerations

Stacked antennas can have higher gain because the structure is thicker. But a thick antenna is not suitable for small devices. To fit antennas into compact devices, engineers always recommend a thin configuration. The majority of the time, a thin air gap is used as a layer between the top and bottom radiators. Because of the thin layer and low permittivity of air, the losses in the structure are minimal while the radiated energy is maximum. Even though the radiation properties and bandwidth remain excellent. However, this configuration is mechanically unstable and weak. Moreover, fabricating such a system with a very thin air layer is challenging. Furthermore, managing the proper positions of both radiators becomes more challenging at mm-wave frequencies. In such frequencies, a fraction of a millimeter can significantly impact the antenna performance. Therefore, stacked antennas require careful manufacturing.

6.4 Parasitic-element Techniques for Bandwidth Enhancement

Parasitic-element techniques are a simple and effective way to increase bandwidth in microstrip antennas. A normal microstrip patch antenna has a narrow bandwidth. It supports only one main resonance. This is not enough for modern systems that need fast data and a wide frequency range. In parasitic-element designs, one or more metal patches are placed close to the main radiator. These patches are not directly fed. They collect energy from the main radiator through electromagnetic coupling. Because of this coupling, the parasitic elements create new resonances. When these resonances are close to the main resonance, they join and make a wide bandwidth. This method is popular in mm-wave antennas because the structure is still small and easy to make (Chattha et al., 2009; Li et al., 2005; Tran et al., 2018). An example of a parasitic-element bandwidth enhancement technique is shown in Figure 6.2, where two parasitic elements are placed near the main patch antenna.

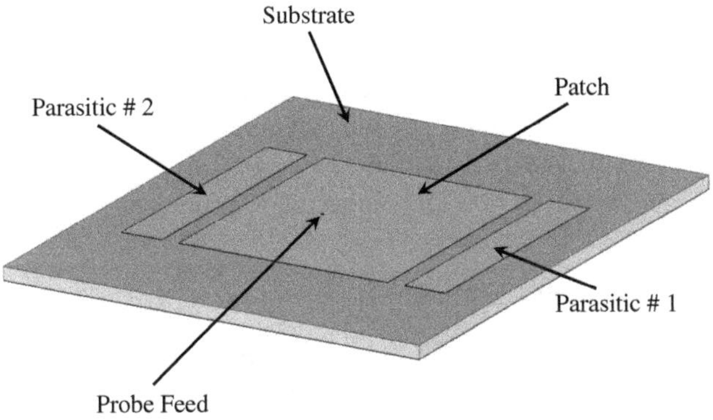

Figure 6.2 An example of a microstrip patch antenna coupled with two parasitic elements. The main patch is excited by the probe feed method.

6.4.1 Basic Idea of Parasitic Elements

A parasitic element receives energy from the main patch without a feed line. This happens through electric and magnetic fields in the space between them. Therefore, the new element has its own natural resonance. It can be tuned by changing its length, shape, or position. When the new resonance lies close to the main resonance, the antenna shows a wide region where impedance is matched. The reflection coefficient remains low for many frequencies. Designers check this with $|S_{11}|$ values. If $|S_{11}| < -10$ dB, the antenna is considered matched. A wider region under this value means a wider bandwidth.

The bandwidth BW relates to the quality factor Q like:

$$\text{BW} \approx \frac{1}{Q}.$$

In the case of a parasitic structure, most of the energy is radiated into the free space, and less energy is stored in the antenna itself. Since Q-factor (Q) is inversely proportional to the radiated energy into the air, it increases the bandwidth. More specifically, antennas having wideband (achieved through parasitic elements) are very useful in high-frequency applications.

6.4.2 Spacing Between Radiator and Parasitic Element

The gap between the primary radiator and secondary parasitic path is vital in antenna performance. This gap should be wisely selected to ensure excellent performance. For example, if the secondary parasitic radiator is kept in very close proximity to the primary radiator, it will even disturb the primary radiator. This is because of the strong coupling between both. This strong coupling disturbs the current distribution in the primary radiator of the antenna. In contrast, if the parasitic radiator is not in close proximity, then the coupling will be very weak. In such scenarios, there is no use of the parasitic structure as it will not play any role in bandwidth enhancement. As a rule of thumb, a gap of less than 0.15λ is always considered a good starting point. This design parameter can be further refined using full-wave electromagnetic simulation tools to achieve optimal performance.

6.4.3 Different Parasitic-element Positions

Parasitic elements can be placed in many ways. Some designs put them beside the main patch. Some put them above or below the main radiator. Some place several surrounding elements around the main patch. Each placement gives a different coupling and band shape. When the element is beside the radiator, the coupling is mostly electric. When the parasitic element is stacked above, the coupling is a mix of electric and magnetic. The designer must test and tune these effects. The final shape of S_{11} curve tells whether the design is good or not.

6.4.4 Directors and Reflectors

In many antenna designs, parasitic elements can act as directors or reflectors. A director is shorter than the main patch and creates a resonance at a higher frequency. A reflector is longer and creates a resonance at a lower frequency. By using one reflector and one director, the designer can create two new resonances. These resonances sit close to the main one. Through their mutual coupling and combined resonant behavior, they enable substantial bandwidth enhancement. This method is simple yet powerful for mm-wave devices.

6.4.5 Multiple Parasitic Elements

In advanced configurations, multiple parasitic elements may be incorporated to further enhance coupling effects and bandwidth performance. Two, three, or more parasitic patches can surround the main radiator. They create many resonances that merge together. However, when the number becomes large, the design becomes very sensitive to small changes. Even a small misalignment or a small change in the gap between the main radiator and the parasitic can significantly alter the results. This becomes more prominent at high-frequency bands, where the main radiator size and parasitic element sizes are extremely small. In such cases, even a fraction of a millimeter can change the matching and resonances significantly. The antenna designer should give careful consideration to this, and such consideration should be more careful at mm-wave frequency bands. In addition, due to resonances from the parasitic element, the radiation pattern changes.

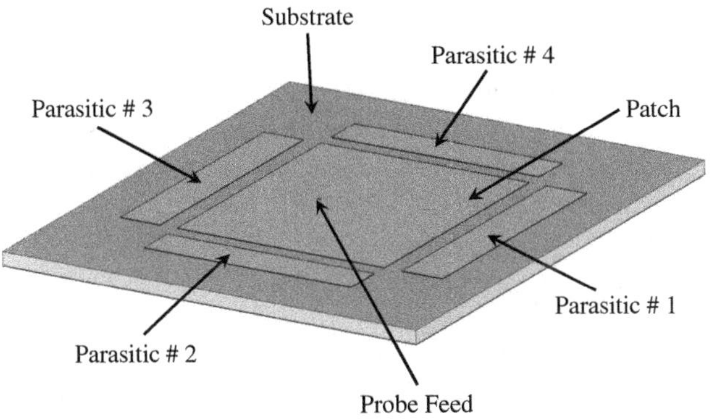

Figure 6.3 An example of a microstrip patch antenna coupled with multiple parasitic elements. The main resonator (patch) is excited by the probe feed method.

The designers must ensure that the primary radiation remains unchanged with introducing parasitic elements. An example of using multiple parasitic elements is shown in Figure 6.3.

6.4.6 Printed Parasitic Structures

Many modern designs use parasitic elements that are printed on the same substrate. These parasitics are found in different shapes, including meandered, spiral, ring, square, triangular, arc, etc. The shape of the parasitic structure mainly depends on the primary resonators. Such structures are placed in the proximity of the main radiator to have efficient coupling. Moreover, the choice of shape also depends on how a second mode can be excited using parasitics. In addition, the choice of shape also depends on whether it should not alter the radiation pattern. The antenna engineers use different approaches, including electromagnetic simulation software, to find a suitable shape. The parasitic should be designed in a manner that the maximum current can easily couple with it without a physical connection. As a result, the reflection curve should be stable (less than −10 dB) across a broad frequency band. If the antenna system encounters short and sharp resonances, there might be a possibility that it will not work efficiently in wireless devices.

6.4.7 Aperture-type Parasitic Elements

In slot antennas, slots are etched in the copper to radiate the electromagnetic waves in free space. The main slot dimensions are conventionally chosen to be $\lambda/2$ or $\lambda/4$. A similar method is applied to use them as parasitic structures. These slots are normally designed within the patch of the antenna. The resonance of the slot and the main radiator resonance merge to form a wide band. They give an extra radiation path under the patch. The coupling between the slot and the patch helps create a second band. This method is useful in mm-wave because ground slots can be made very small but still strong. In addition, slots also help control the direction of the beam, reduce surface waves, and improve efficiency.

6.4.8 Bandwidth and Gain Trade-off

Just like the broad bandwidth of the antenna, radiation properties such as realized gain and total efficiency are also important. For determining the gain and efficiency, special care should be given to placing parasitic resonators. They can enhance or even degrade these parameters. If it acts as a director, then it focuses energy in the direction of the main radiator and increases gain. If they disturb the field and send power to the wrong place, the gain decreases. The designer must check gain across the whole band. If the gain drops at some frequencies, the antenna may not support stable communication. Good design provides both a wide bandwidth and high gain.

6.4.9 Benefits for mm-Wave Systems

Parasitic-element techniques work well for mm-wave because the structures are very small. The coupling is strong at high frequency. This makes it easy to get more than one resonance. The design does not need extra feeding networks. With fewer feeding networks, the losses are low. Lower losses are even more important for high-frequency bands due to their lower immunity to the surroundings. The introduction of printed or slot parasitics can keep the overall size compact. Such a configuration is important for integration in handheld devices, IoT sensor nodes, and other wireless devices.

As parasitics introduce additional resonances, they can support several high-frequency bands. For instance, an antenna system can be designed that covers both 28 and 38 GHz frequency bands, which are important bands for 5G. This can be done by designing a dual-resonance antenna and effectively coupling the resonances to generate a broad bandwidth. Apart from the benefit of having a single wideband antenna covering both bands, it is more resilient to any frequency shift due to the antenna's surroundings. It will ensure resilience in its radiation performance and reliable wireless communication. Due to its resilience to the surrounding environment, it can be integrated into different wireless communication devices. These antennas work well even with RF and electronic parts surrounding them.

6.4.10 Practical Design Challenges

Designing broadband antennas using parasitic structures is a relatively simple approach. However, it has several challenges that need to be considered. These challenges become more important for mm-wave frequency bands, since gaps between the parasitics and the main radiator are extremely small. The main coupling between the primary radiator and parasitic element happens due to the gap. A small variation in the coupling gap seriously impacts antenna resonance and radiation properties. In the case of a patch or monopole antenna as a primary radiator, its radiating edges should be carefully designed. This is due to the fact that they radiate through edges and also couple with parasitic through edges. The printed parasitics are placed on the substrate, and the surface roughness should be considered as it heavily changes the antenna performance. Moreover, the antenna performance gets impacted by the temperature change. This is because the temperature change varies the dielectric properties of the substrate. Subsequently, the antenna resonance moves towards lower or higher frequencies. The antenna designers always choose a substrate that is more resilient to any change in temperature. In this way, the antenna performance remains the same even in different environments and devices. Even though parasitic elements are helpful for getting a wide operating bandwidth, they are highly sensitive. Thus, the designers should always compare the simulated and measured results.

The antenna performance is greatly influenced by its surroundings. For example, the antenna behaves differently when placed in free space and when held in the hands. At the mm-wave frequency band, if a user holds a handheld device, the resonance shifts towards the low-frequency band. Also, the user's hands absorb signals and reduce the antenna's total efficiency and gain. Parasitic elements are useful in reducing such impact by expanding the bandwidth and constructively coupling with the main radiator to enhance gain. However, designing parasitics that concurrently increase bandwidth and gain is challenging. The antenna designer should ensure practical results resemble the simulated ones. An antenna is considered suitable for handheld devices or IoT sensor nodes if its performance is comparable to free space performance.

6.5 Hybrid Techniques

Hybrid techniques improve antenna performance at mm-wave frequencies by providing a wide bandwidth and stable radiation. A hybrid technique is mixing two or more bandwidth enhancement ideas in one design. Each method alone may not give good results. But together, they can support each other. The final antenna can cover a large band and still stay small. Such techniques are useful for applications where concurrent broadband and compact size are necessary (Katehi et al., 1987).

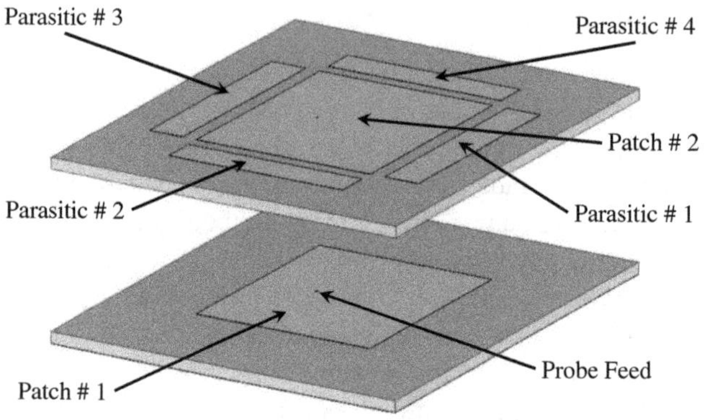

Figure 6.4 An example of a microstrip patch antenna coupled with two parasitic elements and also with a stacked patch. The main patch is excited by the probe feed method.

An example of a hybrid technique for bandwidth improvement is shown in Figure 6.4. It can be seen that the main resonator (patch) is coupled with a secondary stacked patch and also with four parasitic elements. The main patch is excited with the probe-feed technique.

Because of the small wavelengths at mm-wave frequency bands, the antenna size is very small. The natural bandwidth of such small antennas is narrow, and they also have more losses. The antenna engineers try to investigate new methods to resolve these issues. They integrate good features of several techniques to design a compact and broadband antenna. Such an approach mitigates the limitations and improves antenna performance. The main objective behind using such an approach is to design an antenna with stable impedance matching and high total efficiency over a wide frequency range.

Hybrid techniques may consist of combining different approaches within a single design. In this section, we have explained some common hybrid techniques. This will give the reader a better understanding of where to use which specific approach.

6.5.1 Mix of Slot and Patch Antennas

This hybrid technique uses both slot and patch structures at the same time. The patch gives strong radiation, and the slot gives extra resonance. Both together make the bandwidth wider (Venkateshwarlu et al., 2024).

A simple example is a microstrip patch with a slot cut in the middle. The slot creates a second resonance near the first one. When the two resonances merge, the return loss becomes better across a wider band. The input match improves and the antenna can also stay thin.

Sometimes the slot shape changes the current flow, which helps reduce surface waves and keeps more energy in the main beam. Therefore, the antenna gives better gain and supports multiple mm-wave channels.

6.5.2 Mix of Stacked Layers and Parasitic Elements

A stacked antenna places more than one radiating layer on top of each other. A parasitic element is disconnected from the main radiator and also from the feeding section. It is placed in proximity to the main radiator. It is coupled with the main radiator and generates another resonance. When both are coupled tightly, one broadband impedance matching occurs, which includes both excited modes. These parasitic remain on the main layer as the radiating element (Chen et al., 2010).

Another parasitic is designed and placed on a different layer than the first parasitic. This parasitic element also gets some energy from the main radiator and excites a mode to generate resonance. This resonance combines with the main resonance and the resonance generated by the first parasitic. Consequently, a wider bandwidth is obtained without compromising other parameters of the antenna. The second parasitic can be placed very close (adjacent layer) or away (far-end layer) from the first one. This separation gap controls the coupling and ultimately the bandwidth of the antenna.

Another method to couple more than one parasitic with the antenna is by placing the secondary parasitics above or below the antenna in the air. In such cases, the air medium acts as a dielectric, and energy is coupled through it. Considering high-frequency applications such as mm-wave 5G, air gap stacking is efficient. However, it is impractical if the air gap is too large.

6.5.3 Mix of Different Feeds and Radiators

The feed line is where the signal enters the antenna, while the geometry determines how radiation occurs. At mm-wave, a good feed match is very important. So designers mix a strong radiator with a wideband feed (Wu et al., 2018).

A simple microstrip feed may not work well at mm-wave. So a hybrid design may use:

- CPW feed,
- tapered line feed,
- differential feed, and
- aperture coupled feed.

When these feeds are combined with shapes such as slots or tapered patches, the bandwidth increases. A wide feed gives a soft transition, and energy does not reflect. The antenna becomes easy to match across multiple GHz.

6.5.4 Mix of Slots and Parasitic Edges

Some antennas have slots in the radiator and parasitic edges around the main shape. The slot creates one resonance, while the parasitic edge creates another. When both resonances are close, the band becomes wider.

This idea looks simple: designers cut small shapes in the metal layer. They also attach short strips near the radiator. The current flows into all these features. The result is strong and wide radiation.

Parasitic edges also help reduce side radiation. They push more energy to the front. This is good for communication links where the direction must be clear.

6.5.5 Mix with Electromagnetic Band Gap

Electromagnetic band gap (EBG) structures sit near the ground plane. They control surface waves, which can waste energy. They make the gain low. With an EBG, the radiation goes up more. This makes the antenna strong across the band (Nasimuddin et al., 2016).

When we mix an EBG with a wideband patch, the bandwidth becomes larger. The antenna keeps stable radiation even at the edge of the band. This is because the EBG removes unwanted coupling.

6.5.6 Mix of Tuning Stubs and Modified Ground

A tuning stub adds extra inductance or capacitance. A modified ground changes the path of the current. Both help improve the match. When used together, the result is better than using only one (Jeong et al., 2012).

For example, if the antenna has a poor match at the upper band, a ground slot may fix it. If the lower band is weak, a stub can help. When designers tune both, they get a good match across the full mm-wave range.

6.5.7 Advantages of Hybrid Techniques

Hybrid techniques provide multiple performance benefits, including improved bandwidth, gain stability, and radiation efficiency through combined design mechanisms.

When two or more ideas combine, they can complement each other and overcome individual limitations. The slot helps improve matching, the parasitic element helps increase gain, the stacked patch helps create more resonances, and the wide feed helps improve energy transfer. This is why hybrid methods are popular today.

Hybrid techniques also help in real products. Phones, cars, and radar units want small antennas. The goal is to design a single antenna that operates over multiple frequency bands, and hybrid design techniques help achieve this objective.

6.5.8 Design Challenges

Even with many benefits, there are problems with hybrid methods. When we add many parts, the design becomes complex. It is hard to predict the result. The antenna may act in a different way at mm-wave.

Fabrication is also a challenge. Small errors can change the result a lot. For example, a small shift in parasitic distance can move the resonance many GHz. Designers must rely on high-precision manufacturing equipment during fabrication.

Simulation time is also long because the model has many layers and shapes, requiring more computational effort, which makes the design process longer.

6.5.9 Practical Examples

Many antennas today use hybrid ideas. A phone mm-wave antenna may use the following:

- a patch,
- a slot inside the patch,
- parasitic sheets on the side, and
- a special CPW feed.

All of this fits within a few millimeters and provides a wide bandwidth, such as 26–40 GHz, which supports 5G applications. A radar antenna in a car may use stacked parasitic layers to provide high gain and wide coverage at 77 GHz.

These examples show that hybrid methods are not just theoretical; they work in real-life applications.

6.6 Conclusion

This chapter discussed the importance of wideband antennas for mm-wave communication systems and explained why wide bandwidth is essential for modern wireless technologies. As mm-wave systems aim to support high data rates, low latency, and advanced features such as beamforming and large MIMO, antennas must operate over broad frequency ranges to maintain reliable performance. The chapter reviewed several practical bandwidth enhancement techniques, including geometric modification, stacked microstrip structures, parasitic elements, and hybrid design approaches. Each method introduces additional resonances or improves impedance matching to extend the usable frequency band. It was also highlighted that careful design of coupling, feeding structures, and materials is necessary to achieve stable wideband performance at mm-wave frequencies. Overall, combining different techniques provides the most effective solution for designing compact, efficient, and broadband antennas suitable for next-generation wireless systems.

References

H.F. AbuTarboush, H.S. Al-Raweshidy, and R. Nilavalan. Bandwidth enhancement for microstrip patch antenna using stacked patch and slot. In *2009 IEEE International Workshop on Antenna Technology*, pages 1–4, 2009. https://doi.org/10.1109/IWAT.2009.4906874.

Segun Akinola, Iddi Hashimu, and Ghanshyam Singh. Gain and bandwidth enhancement techniques of microstrip antenna: A technical review. In *2019 International Conference on Computational Intelligence and Knowledge Economy (ICCIKE)*, pages 175–180, 2019. https://doi.org/10.1109/ICCIKE47802.2019.9004278.

Warsha Balani, Mrinal Sarvagya, Tanweer Ali, Manohara Pai M.M., Jaume Anguera, Aurora Andujar, and Saumya Das. Design techniques of super-wideband antenna–existing and future prospective. *IEEE Access*, 7:141241–141257, 2019. https://doi.org/10.1109/ACCESS.2019.2943655.

H.T. Chattha, Y. Huang, Y. Lu, and X. Zhu. Further bandwidth enhancement of PIFA by adding a parasitic element. In *2009 Loughborough Antennas & Propagation Conference*, pages 213–216, 2009. https://doi.org/10.1109/LAPC.2009.5352441.

Yikai Chen, Shiwen Yang, and Zaiping Nie. Bandwidth enhancement method for low profile E-shaped microstrip patch antennas. *IEEE Transactions on Antennas and Propagation*, 58(7):2442–2447, 2010. https://doi.org/10.1109/TAP.2010.2048850.

Zhi Ning Chen. Development of ultra-wideband antennas. *Journal of Electromagnetic Engineering and Science*, 13(2):63–73, 2013.

Renato Cicchetti, Emanuela Miozzi, and Orlandino Testa. Wideband and UWB antennas for wireless applications: A comprehensive review. *International Journal of Antennas and Propagation*, 2017(1):2390808, 2017.

Gye-Taek Jeong, Woo-Su Kim, and Kyung-Sup Kwak. Dual-band Wi-Fi antenna with a ground stub for bandwidth enhancement. *IEEE Antennas and Wireless Propagation Letters*, 11:1036–1039, 2012. https://doi.org/10.1109/LAWP.2012.2214755.

P. Katehi, N. Alexopoulos, and I. Hsia. A bandwidth enhancement method for microstrip antennas. *IEEE Transactions on Antennas and Propagation*, 35(1):5–12, 1987. https://doi.org/10.1109/TAP.1987.1143970.

B.A. Kramer, C.-C. Chen, M. Lee, and J.L. Volakis. Fundamental limits and design guidelines for miniaturizing ultra-wideband antennas. *IEEE Antennas and Propagation Magazine*, 51(4):57–69, 2009. https://doi.org/10.1109/MAP.2009.5338684.

Alok Kumar, Nancy Gupta, and PC Gautam. Gain and bandwidth enhancement techniques in microstrip patch antennas—a review. *International Journal of Computer Applications*, 148(7):9–14, 2016.

Gurpreet Kumar and Rajeev Kumar. A survey on planar ultra-wideband antennas with band notch characteristics: Principle, design, and applications. *AEU—International Journal of Electronics and Communications*, 109:76–98, 2019.

Seong H Lee, Jong K Park, and Jung N Lee. A novel CPW-fed ultra-wideband antenna design. *Microwave and Optical Technology Letters*, 44(5):393–396, 2005.

RongLin Li, G. DeJean, J. Laskar, and M.M. Tentzeris. Investigation of circularly polarized loop antennas with a parasitic element for bandwidth enhancement. *IEEE Transactions on Antennas and Propagation*, 53(12):3930–3939, 2005. https://doi.org/10.1109/TAP.2005.859917.

Brijesh Mishra, Ramesh Kumar Verma, Yashwanth N, and Rakesh Kumar Singh. A review on microstrip patch antenna parameters of different geometry and bandwidth enhancement techniques. *International Journal of Microwave and Wireless Technologies*, 14(5):652–673, 2022.

A.A. Mohammad, H. Subhi, A.K. Ahmad, and S.M. Juma. Bandwidth enhancement of stacked rectangular microstrip patch antenna. In *2006 7th International Symposium on Antennas, Propagation and EM Theory*, pages 1–13, 2006. https://doi.org/10.1109/ISAPE.2006.353389.

Tapan Nahar and Sanyog Rawat. Survey of various bandwidth enhancement techniques used for 5G antennas. *International Journal of Microwave and Wireless Technologies*, 14(2):204–224, 2022.

N. Nasimuddin, Zhi Ning Chen, and Xianming Qing. Bandwidth enhancement of a single-feed circularly polarized antenna using a metasurface: Metamaterial-based wideband CP rectangular microstrip antenna. *IEEE Antennas and Propagation Magazine*, 58(2):39–46, 2016. https://doi.org/10.1109/MAP.2016.2520257.

Elsuke Nishiyama, M Aikawa, and S Egashira. Stacked microstrip antenna for wideband and high gain. *IEE Proceedings-Microwaves, Antennas and Propagation*, 151(2):143–148, 2004.

Tale Saeidi, Idris Ismail, Wong Peng Wen, Adam RH Alhawari, and Ahmad Mohammadi. Ultra-wideband antennas for wireless communication applications. *International Journal of Antennas and Propagation*, 2019(1):7918765, 2019.

Yuan Shen and Moe Z. Win. On the accuracy of localization systems using wideband antenna arrays. *IEEE Transactions on Communications*, 58(1):270–280, 2010. https://doi.org/10.1109/TCOMM.2010.01.080141.

Huy Hung Tran, Nghia Nguyen-Trong, Truong Khang Nguyen, and Amin M. Abbosh. Bandwidth enhancement utilizing bias circuit as parasitic elements in a reconfigurable circularly polarized antenna. *IEEE Antennas and Wireless Propagation Letters*, 17(8):1533–1537, 2018. https://doi.org/10.1109/LAWP.2018.2853193.

G. Venkateshwarlu, G. Vallathan, Senthilkumar Meyyappan, and Mohammed Abdul Nasar. Microstrip patch slot antenna with enhanced bandwidth and multiple parasitic patches. In *2024 Third International Conference on Intelligent Techniques in Control, Optimization and Signal Processing (INCOS)*, pages 1–6, 2024. https://doi.org/10.1109/INCOS59338.2024.10527755.

Qiong-Sen Wu, Xiao Zhang, and Lei Zhu. A wideband circularly polarized patch antenna with enhanced axial ratio bandwidth via co-design of feeding network. *IEEE Transactions on Antennas and Propagation*, 66(10):4996–5003, 2018. https://doi.org/10.1109/TAP.2018.2851616.

K.Y. Yazdandoost and R. Kohno. Ultra wideband antenna. *IEEE Communications Magazine*, 42(6):S29–S32, 2004. https://doi.org/10.1109/MCOM.2004.1304230.

7

Millimeter-wave Planar Antenna Arrays

7.1 Introduction

Applications like fifth-generation (5G) communications, radar imaging, driverless cars, satellite communications, and high-speed wireless links can all benefit from the use of millimeter-wave (mm-wave) technology. But there are additional difficulties when working at mm-wave frequencies. Because of the extremely small wavelength (measured in millimeters) at these high frequencies, antennas must be designed with higher precision and compact dimensions. High-gain and effective antennas are necessary to guarantee dependable communication because path loss and atmospheric attenuation increase with frequency (Ghosh et al., 2022; Zhao et al., 2021).

To overcome these limitations, antenna arrays have been widely used. An antenna array is a group of individual antenna elements arranged in a specific pattern and excited with specific amplitudes and phases. By properly designing the array, it is possible to increase the gain, control the beam direction, and improve the radiation performance (Bang and Choi, 2020). Among various types of antenna arrays, planar antenna arrays have become very popular at mm-wave frequencies because they are flat, compact, lightweight, and easy to integrate with modern printed circuit technologies (Maggi et al., 2022; Naqvi and Lim, 2018; Sadhu et al., 2019).

7.1.1 What Is a Planar Antenna Array?

A planar antenna array is a two-dimensional arrangement of multiple antenna elements placed on a flat surface or substrate. Each element in the array radiates electromagnetic waves, and when these waves combine in space, they form a specific radiation pattern. By adjusting the phase or amplitude of each element, the main lobe of the radiation pattern can be steered in different directions, and the side lobes can be reduced.

Planar arrays can be made using different antenna types such as microstrip patches, slots, dipoles, or horns. However, microstrip patch-based arrays are the most commonly used at mm-wave frequencies because of their planar structure, low profile, and compatibility with printed circuit boards and monolithic microwave integrated circuit fabrication processes. This makes them suitable for low-cost mass production.

The design of a planar array usually involves choosing the type of element, its size and shape, the inter-element spacing, the feeding network, and the overall array geometry (for example, rectangular, circular, or triangular arrangements). Each of these factors plays a key role in determining the final performance of the antenna in terms of gain, bandwidth, efficiency, and beam steering capability. An example of a 2 × 2 planar patch array is shown in Figure 7.1. This antenna array is designed in CST Microwave Studio, which operates between 26.5 and 27 GHz, as shown in Figure 7.2. It provides a realized gain of 3.1 dBi in the band of interest. The radiation pattern of the antenna array is shown in Figure 7.3.

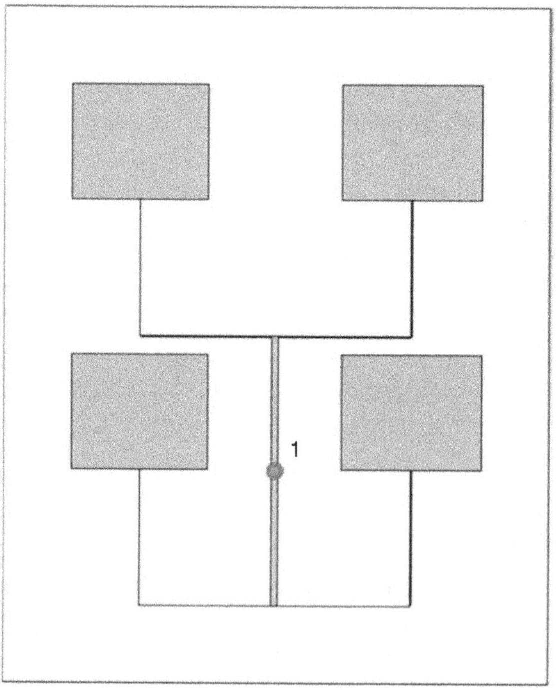

Figure 7.1 An example of 2 × 2 planar patch array.

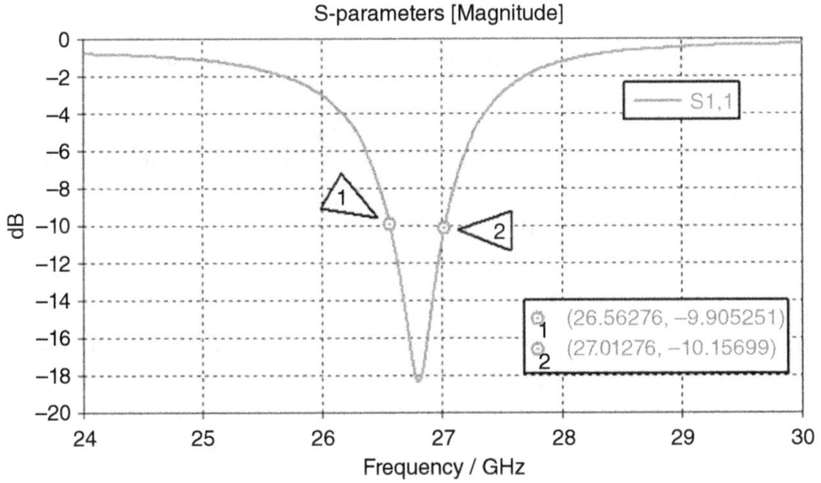

Figure 7.2 S-parameters of the 2 × 2 planar patch array.

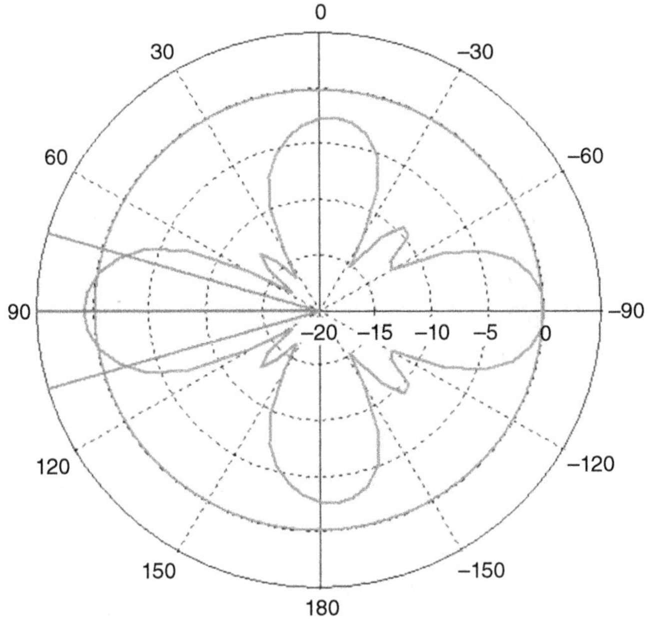

Figure 7.3 Radiation pattern of the 2 × 2 planar patch array.

7.1.2 Importance of mm-Wave Planar Arrays

At mm-wave frequencies, the wavelength becomes very small; for example, it is approximately 10 mm at 30 GHz and 3 mm at 100 GHz. Because of this, antenna elements can be made very compact, and many elements can be placed in a small area. This allows the creation of large arrays with high directivity and beamforming capability without increasing the overall antenna size.

Among the many advantages of planar array antennas, some of the few are listed as follows:

1. **High Gain:** As the path loss at mm-wave frequencies is high, a single antenna element often cannot provide sufficient gain. By doubling the number of elements, the antenna gain can be increased by 3 dBi. As antenna arrays contain multiple identical or similar radiating elements, increasing the number of elements can improve the overall antenna gain. At high frequencies, high gain is desirable to compensate for the free-space path losses. With high gain numbers, the link margin can be improved.
2. **Beam Steering and Beamforming:** The demand for next-generation (such as 5G) wireless networks is exponentially increasing. These networks use massive multiple-input multiple-output (MIMO) transceivers. Thus, beam steering and beamforming are a vital part of them. Specifically, beam steering is important for tilting the beam toward the intended user. In planar arrays, these functionalities can be achieved electronically using diodes or micro-electro-mechanical systems (MEMS). In fact, electronic switches can be used to control or change the phase of an individual element in the array system. This useful functionality of planar arrays reduces system interference and increases its capacity.
3. **Compact and Low-profile Design:** Since mm-wave wavelengths are small, planar arrays can be designed to be very compact and lightweight. This is very useful for mobile devices, vehicles, and satellite systems where space is limited.
4. **Integration with Circuits:** Planar arrays can be fabricated using the same technology as radio frequency (RF) and microwave circuits. Therefore, integrating planar arrays with other RF circuitry becomes easier and simpler. This approach is useful in system-in-package applications where all components are integrated into a single substrate.
5. **Multiple Applications:** mm-Wave planar arrays are not only used in 5G communication but also in radar sensors for vehicles, imaging systems for security and medical applications, and satellite communication links.

Apart from the attributes mentioned previously, there are a number of other important functionalities. Thus, planar antenna arrays have been widely adopted in next-generation wireless networks.

7.1.3 Design Considerations for mm-Wave Planar Arrays

Designing planar arrays at mm-wave frequencies requires careful consideration of several factors. These include element design, array spacing, mutual coupling, feed network, fabrication tolerance, and material losses (Pinchera, 2020). Key design considerations for mm-wave planar antenna arrays are shown in Figure 7.4.

- **Element Design:** To develop an antenna array, an efficient unit element should be designed. The element geometry, shape, operating frequency, bandwidth, gain, and efficiency should be optimized before integrating it into an array configuration. For planar arrays, the common design topologies include monopole antennas, slot antennas, and patch antennas. Any of the aforementioned topologies can be selected based on the physical and radiation requirements.
- **Element Spacing:** When a unit element is designed, it performs according to the requirements. Then, it is integrated into the array configuration. These optimized elements are arranged in a precise manner. The spacing between the antenna array elements is very important in maintaining good radiation performance. For a larger space between the elements, the antenna array may suffer from grating lobes. The performance may suffer from mutual coupling between elements if it is too small. To counteract these effects, element spacing is usually maintained at about half a wavelength.

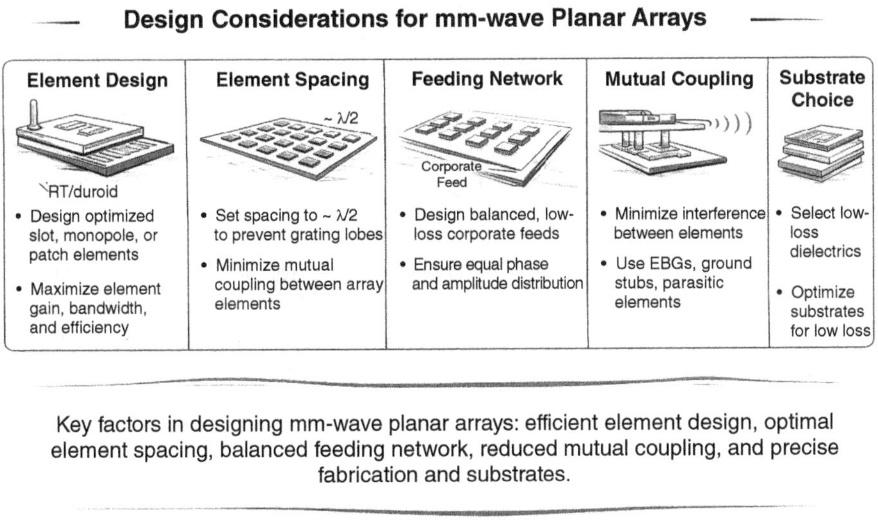

Figure 7.4 Key design considerations for millimeter-wave planar antenna arrays.

- **Feeding Network:** All the elements receive power from the feed network in the proper phase and amplitude. High losses in transmission lines and connectors make it difficult to design a low-loss, balanced feeding network at mm-wave frequencies. To overcome these challenges, several feeding network designs have been presented in the literature. The most common ones are series feeding, parallel feeding, and corporate feeding. In addition, substrate integrated waveguide (SIW)-based series and parallel feedings are also very common and useful.
- **Mutual Coupling:** In antenna arrays, the elements are arranged in close proximity. The EM fields originating from one element can affect the adjacent elements. Such a phenomenon is called mutual coupling. As previously mentioned, the performance may suffer from mutual coupling between elements if the spacing is too small. This undesired mutual coupling can decrease the overall performance in terms of radiation efficiency and gain. To resolve the mutual coupling problem, several methods have been proposed in the literature. Among them, the common ones are defective ground, using electromagnetic band gaps (EBGs), using stubs, and lossy parasitic structures.
- **Fabrication Tolerances:** The geometry of antenna arrays working at mm-wave frequencies is very small. If the structure is based on a $\lambda/4$ monopole radiator, it will have very small copper traces. A small discrepancy in the fabricated trace will lead to a high-frequency shift. Therefore, very accurate circuits should be designed for these antennas.
- **Substrate Choice:** Substrates are always used to print planar structures. The electrical characteristics of the substrate have a significant effect on antenna performance. Low-loss materials are always preferred when designing antennas.

7.1.4 Applications of Planar Arrays

7.1.4.1 5G and Beyond Communication Systems

The mm-wave frequency bands are widely adopted for next-generation wireless communication systems (5G and beyond). Antenna arrays in planar form are used in base stations and mobile devices to provide directional beams and high gain. In the future, 6G systems are expected to use even higher frequencies, such as the sub-terahertz range, where planar arrays will remain a key enabling technology.

7.1.4.2 Automotive Radar

In modern vehicles, radar systems operating at 77 GHz are used for adaptive cruise control, collision avoidance, and parking assistance. In these systems, planar antenna arrays allow compact radar sensors to detect objects at long ranges and with high angular resolution. Because of their flat shape, they can be easily mounted on the car body or behind plastic panels without affecting the appearance.

7.1.4.3 Satellite and Space Communication

Satellite communication at mm-wave frequencies allows higher data throughput while using smaller antennas. Planar arrays are particularly useful for satellite terminals, as they can form multiple beams and electronically track satellites without mechanical movement. Moreover, they are useful for nonstop beam scanning and screening for low earth orbit (LEO) satellites.

7.1.4.4 Imaging and Sensing Systems

One of the key applications of mm-wave frequencies is imaging. This capability is critical for defense applications (weapon screening), enhancing security, and improving medical imaging. By using recorded reflections from targeted objects, planar arrays can be used for generating high-quality images. More specifically, they have a very small size due to their small wavelength. Thus, they have become highly attractive for use in compact and portable imaging systems.

7.1.5 Challenges in mm-Wave Planar Array Design

Planar arrays operating at mm-wave frequencies have many advantages and enable numerous applications. With many advantages, however, they have several challenges to overcome. Recently, several approaches have been proposed to address these challenges.

- **High Losses:** Planar arrays have metal deposition on both top and bottom surfaces of the substrate. Both substrate and copper (or any other metallic structure) contribute to the total losses of the system. More specifically, these losses increase by several factors at mm-wave frequencies.
- **Fabrication Precision:** Planar arrays for mm-wave applications have a very small footprint. They are normally in mm range because of their extremely small wavelengths. Fabricating such a small structure requires very precise and advanced techniques.
- **Complex Feeding Networks:** The elements of the planar arrays are connected to each other using a feeding network. The feeding networks have intrinsic losses. Moreover, the design of a feeding network for a large number of elements is complex, bulky, and prone to losses.
- **Thermal Management:** High-power operation at mm-wave bands can cause heating issues, especially in active phased arrays.

- **Cost:** Conventional PCB fabrication methods can be used to design mm-wave planar antenna arrays. Due to the small geometry of these antennas, designing them in a precise manner is challenging. Designing these antennas using advanced methods increases the cost.

Ongoing research focuses on new materials, improved feeding methods, and integrated designs that combine antennas and circuits to overcome these limitations.

7.1.6 Summary

An overview of mm-wave planar antenna arrays has been given in this chapter. First, we have discussed planar arrays and why they are important at mm-wave frequencies. Planar antenna arrays have narrow beams, directive radiation patterns, high gain, and directivity. They can be easily used in electronic beam scanning and beam steering. They are easy to integrate with other planar RF circuitry. Then, the theory of planar arrays is discussed along with some useful equations. We have also stressed the challenges and potential applications. Moreover, design considerations and best practices are also given in detail.

7.2 Series-fed Antenna Arrays

7.2.1 Introduction

Among applications in different domains, the array structures are used to improve the antenna gain. Antenna gain is important for link budget analysis. For instance, higher gain values are better for receiver sensitivity. Moreover, they are used for beam steering at a certain elevation or azimuth. This application is considered the backbone of future generation technologies. Among different array types, the series-fed array represents a simple and cost-effective configuration. This section of the book will cover topics related to series-fed arrays. More specifically, topics such as series-fed array theory, design equations, limitations, and practical applications are covered in detail (Chen and Otto, 2009; Jones et al., 1982; Khalily et al., 2016; Pozar and Schaubert, 1993).

A series-fed array has several antenna elements connected in series. The start of the transmission line is connected to the source, followed by each element connected in series. In fact, the start of the transmission line for one element is the end of the transmission line for the following element, and so on. Similarly, the current flows in the element near the source, followed by the others, and reaches the end of the array. This feeding is different from conventional array schemes, where antenna elements are connected in parallel. Series-fed arrays are particularly useful at mm-wave frequencies because they simplify the feeding network and reduce the complexity of the design. An example of a series-fed array is shown in Figure 7.5. It contains four patch antennas connected in series. The S-parameters of the series-fed array are shown in Figure 7.6.

7.2.2 Basic Concept of Series-fed Arrays

A series-fed array has several antenna elements connected in series. The transmission line is the medium connecting all elements together. The source is applied at

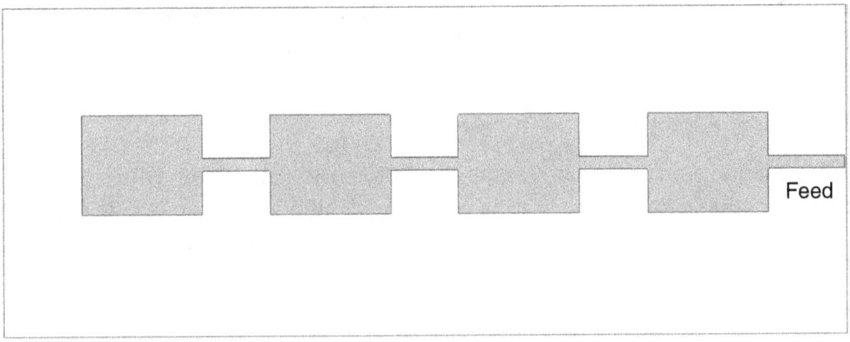

Figure 7.5 Series-fed planar array.

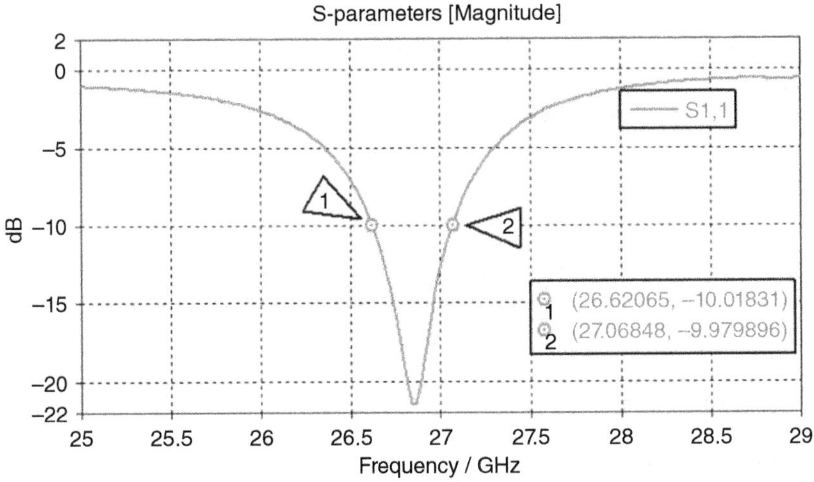

Figure 7.6 S-parameters of the series-fed four-element planar array.

one end of the transmission line and is distributed to the elements sequentially. More specifically, the current travels along the elements through the interconnected transmission line. The spacing between antenna elements is critical for effective coupling and efficient radiation.

7.2.2.1 Array Geometry

Consider a linear-fed planar array with a total element of N. All elements are placed at an equal distance of d. All elements have similar geometry, size, and shape. They also operate at the same frequency band. The start of the transmission line is connected to the source, followed by each element connected in series. The excitation current can be denoted as I_n, when it reaches the last element (nth element). Because of radiation from earlier elements, the current distribution along a lossless line should ideally drop gradually throughout the line.

7.2.3 Array Factor of Series-fed Arrays

A fundamental concept in array antenna theory is the array factor, which describes how the arrangement and excitation of an array's elements influence its radiation pattern. The array factor of an antenna array, where all elements are fed in series, can be computed as follows:

$$\mathrm{AF}(\theta) = \sum_{n=0}^{N-1} I_n e^{jn(kd\cos\theta+\beta)}, \tag{7.1}$$

where:

- I_n is the amount of current present on the last element of the array (nth element)
- $k = 2\pi/\lambda$ is the wave number
- d is the inter-element spacing
- θ is the array axis angle
- β is the phase progression

The above equation can be written in a very simple form when the current distribution over the array is considered as uniform ($I_n = I_0$). The updated equation based on the aforementioned condition is written as follows:

$$\mathrm{AF}(\theta) = I_0 \frac{1 - e^{jN(kd\cos\theta+\beta)}}{1 - e^{j(kd\cos\theta+\beta)}}. \tag{7.2}$$

The above equation is simple and widely used for the main lobe direction, side lobes, and pattern beamwidth.

7.2.3.1 Beam Direction and Phase Control

The planar antenna array's primary beam is produced when the denominator's exponent is equal to zero, that is,

$$kd \cos \theta + \beta = 2\pi m, \quad m = 0, \pm 1, \pm 2, \dots . \tag{7.3}$$

Solving for θ gives the main beam angle:

$$\theta_m = \cos^{-1}\left(\frac{2\pi m - \beta}{kd}\right). \tag{7.4}$$

Without physically moving the antenna, the beam can be electronically guided by varying the phase progression β. This makes it very useful for many applications where beam steering is required without physically moving the antenna.

7.2.4 Design of Series-fed Arrays

The design of a series-fed antenna array is not straightforward and requires careful consideration of factors, such as spacing between the antennas, coupling between the radiators, and tapering of amplitudes. Some of the important points are discussed as follows:

7.2.4.1 Element Spacing

In many applications of wireless communications, antennas are required to transfer or receive information using directive beams. The more effective way to design such an antenna is to have fewer unwanted lobes (grating lobes). These lobes can be reduced or totally suppressed by keeping $\lambda/2$ distance between the adjacent radiators.

7.2.4.2 Amplitude Tapering

Tapering the current along the series line helps reduce sidelobe levels. Common tapers include:

- **Uniform Taper:** $I_n = I_0$
- **Binomial Taper:** Reduces side lobes to almost zero
- **Chebyshev Taper:** Optimizes trade-off between beamwidth and side lobes

7.2.4.3 Coupling Mechanism

Series-fed arrays often use edge coupling, aperture coupling, or proximity coupling between elements. The coupling must be strong enough to transfer energy efficiently but weak enough to maintain proper phase distribution.

7.2.5 Transmission Line Model

Series-fed arrays can be modeled as transmission lines with shunt radiating elements. The voltage and current along the line satisfy the following:

$$\frac{dV(z)}{dz} = -j\beta V(z) - R_r I(z), \tag{7.5}$$

$$\frac{dI(z)}{dz} = -j\beta I(z) - G_r V(z), \tag{7.6}$$

where:

- R_r is the radiation resistance per unit length
- G_r is the radiation conductance per unit length
- β is the propagation constant along the line

This model helps analyze the current attenuation along the line and predict the radiation pattern.

7.2.6 Radiation Pattern Analysis

The total radiation pattern of a series-fed array is the product of its element pattern $E_e(\theta)$ and the array factor $\mathrm{AF}(\theta)$:

$$E_{\mathrm{total}}(\theta) = E_e(\theta) \cdot \mathrm{AF}(\theta). \tag{7.7}$$

For microstrip patch elements, $E_e(\theta)$ is typically broad in the array plane. The array factor shapes the beam and controls the directivity.

7.2.7 Advantages of Series-fed Arrays

Series-fed array configurations provide multiple benefits:

- Simple feeding network
- Lightweight and low profile
- Suitable for mm-wave frequencies
- Easy integration with printed circuits

7.2.8 Limitations

Despite their advantages, series-fed arrays exhibit several limitations:

- The elements of the array system have nonuniform current distribution along the series line. In fact, the surface currents are unequal in all elements.
- It suffers from limited bandwidth.

- In large arrays, the feed line losses are large due to extended lengths. Thus, they have reduced total and radiation efficiency for large arrays.
- For getting exact beam steering at a certain angle, logical phase design is vital. However, maintaining a proper phase is challenging for large arrays.

7.2.9 Practical Applications

Series-fed arrays are applicable in several practical scenarios:

- 5G and beyond communication systems
- Automotive radar sensors
- Satellite terminals
- mm-wave imaging systems

7.2.10 Example: Series-fed Microstrip Array

Consider a four-element series-fed microstrip array with $\lambda/2$ spacing at 30 GHz. The array factor is:

$$\mathrm{AF}(\theta) = \sum_{n=0}^{3} e^{jnkd\cos\theta} = 1 + e^{jkd\cos\theta} + e^{j2kd\cos\theta} + e^{j3kd\cos\theta} \tag{7.8}$$

Assuming $d = \lambda/2$, the main lobe occurs at $\theta = 90°$, and the first side lobes are at $\theta \approx 63°$ and $117°$. The total radiation pattern can be obtained by multiplying this with the patch element pattern.

7.2.11 Summary

Series-fed arrays provide a simple and effective solution for high-frequency antenna design. By understanding the array factor, transmission line model, and current distribution, engineers can design arrays with desired beam direction, gain, and sidelobe performance. Despite some limitations, series-fed arrays remain widely used in mm-wave communication, radar, and sensing systems.

7.3 Parallel-fed Arrays

7.3.1 Introduction

Parallel-fed arrays are one of the most common types of antenna feeding networks used in modern array design. Transmission lines and power dividers are used in such a configuration to connect all elements in a parallel manner (Foglia Manzillo et al., 2015; Hou et al., 2019; Ji et al., 2022). This array topology is more complex in nature compared to series-fed arrays. In a series-fed topology, all radiating elements are connected in a straight line. In fact, each radiating structure is fed in a series manner. In parallel-fed arrays, radiating antennas are connected in parallel. Every radiating part receives a power signal of equal magnitude but of different phase angle. The phase angle between the radiating elements can be controlled through the length of transmission lines. A proper phase difference between the array elements is useful for beam tilting applications. Thus, such antennas are very useful for beam steering applications because of the control over the phase angles (Hirokawa et al., 1992; Huang, 1992; Wei et al., 2023). A typical planar patch array is shown in Figure 7.7. The S-parameters of this parallel-fed patch array are shown in Figure 7.8. The radiation pattern is shown in Figure 7.9. It has a realized gain of better than 13.2 dBi in the band of interest.

In many mm-wave high-frequency devices, these antenna arrays are used. Some of the common applications are 5G communication, radar communication, and satellite communication. The main reason is that all elements of the array have the same power level. Moreover, the phase angles between the radiating parts can be controlled. This flexibility makes it more suitable for mm-wave applications compared to the series-fed array.

7.3.2 Parallel Feeding Basics

This architecture divides the input power into several branches, with each branch feeding one or more antenna elements. This type of feeding will excite all elements with the same frequency. More specifically, the phase distribution and the amplitude of the antenna elements should be properly managed. Similar to a tree, the network starts with a single source and uses smaller pathways to distribute power among its constituent elements (Hirokawa and Ando, 2000; Li et al., 2019; Rawat et al., 2016).

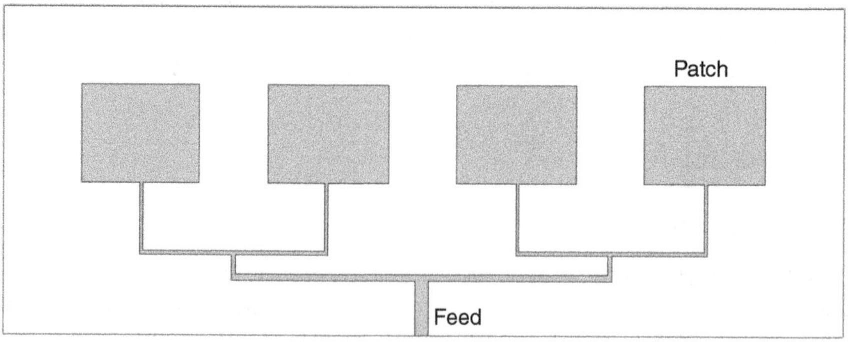

Figure 7.7 Parallel-fed planar array.

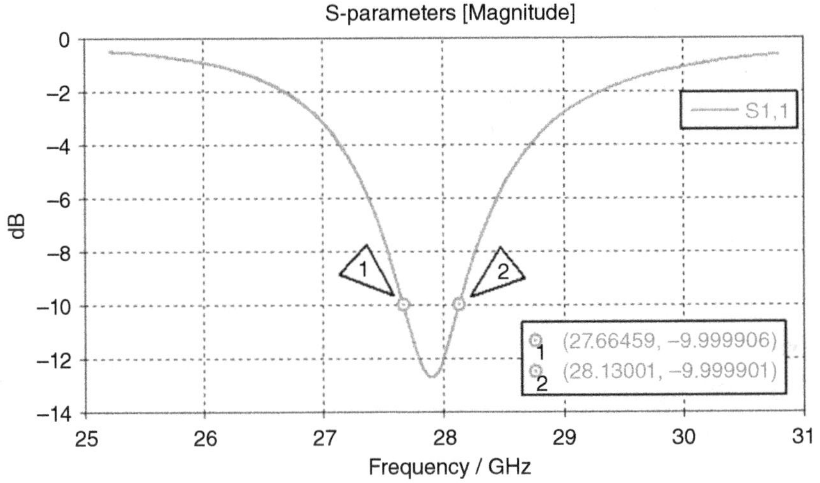

Figure 7.8 S-parameters of the parallel-fed planar array.

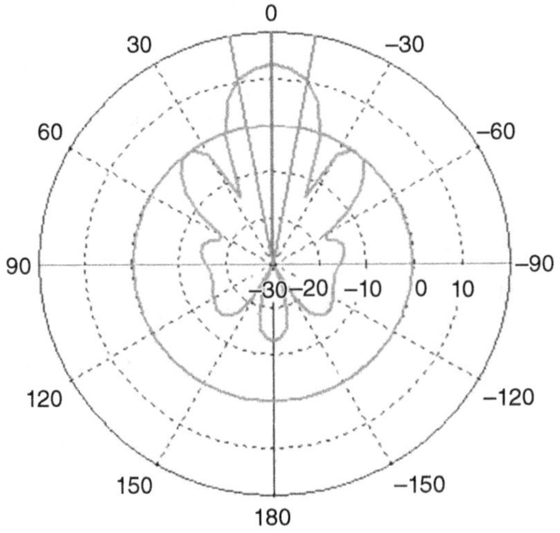

Figure 7.9 Radiation pattern of the parallel-fed planar array.

Mathematically, the current at each antenna element can be written as:

$$I_n = A_n e^{j\phi_n} \tag{7.9}$$

where A_n is the amplitude of the current at the nth element and ϕ_n is its phase. In a uniform parallel-fed array, all amplitudes are equal ($A_n = A$), and the phase is controlled to achieve the desired beam direction.

7.3.3 Array Factor and Radiation Pattern

The radiation pattern of any antenna array depends on the array factor (AF). The array factor describes how the fields from individual elements combine in space.

For a linear array with N elements spaced by distance d, the array factor can be expressed as:

$$\mathrm{AF}(\theta) = \sum_{n=1}^{N} A_n e^{j[k(n-1)d \sin\theta + \phi_n]}, \tag{7.10}$$

where $k = 2\pi/\lambda$ is the wave number and λ is the wavelength.

If all elements are fed with equal amplitude and a progressive phase shift β, the array factor becomes:

$$\mathrm{AF}(\theta) = A \frac{\sin\left(\frac{N}{2}(kd \sin\theta + \beta)\right)}{\sin\left(\frac{1}{2}(kd \sin\theta + \beta)\right)}. \tag{7.11}$$

The main beam direction is found by setting the term in the denominator equal to zero, leading to the condition:

$$kd \sin\theta_0 + \beta = 0 \tag{7.12}$$

Hence, the beam direction θ_0 can be controlled by adjusting the phase shift β in the feed network.

7.3.4 Power Division and Phase Control

In a parallel-fed array, the design of the feeding network plays a key role in how the input power is divided among the elements. The simplest power divider splits the input signal equally into two or more branches. Each branch must be designed to maintain impedance matching to avoid reflections.

The input impedance seen by the source should remain constant and is given by:

$$Z_{\mathrm{in}} = \frac{Z_0}{N}, \tag{7.13}$$

for N identical elements connected in parallel, where Z_0 is the characteristic impedance of each branch.

To maintain good matching, impedance transformers or quarter-wave transformers are often used. For a transmission line of characteristic impedance Z_1 connected between two impedances Z_a and Z_b, the matching condition is:

$$Z_1 = \sqrt{Z_a Z_b}. \tag{7.14}$$

Phase control is another important part of the design. Each feed line is adjusted in length to introduce the desired phase shift between adjacent elements. The phase shift ϕ due to a transmission line of length l is given by:

$$\phi = \beta l = \frac{2\pi l}{\lambda}. \tag{7.15}$$

Based on the above equation, it can be observed that the phase of the overall system can be controlled by varying the length l. This type of adjustment is beneficial for beam steering in the intended direction.

7.3.5 Types of Parallel Feeding Networks

Parallel feeding networks can be constructed using different approaches. The topology and approach depend on the frequency band and its application. There are two common feeding networks based on parallel feeding topology.

7.3.5.1 Corporate Feed Network

In a corporate feed network configuration, the power is allocated as a binary tree in a symmetrical corporate feed. Each divider splits the energy into two equal parts until all the elements are fed. This network has equal amplitude and phase at all outputs, making it ideal for high-performance phased arrays. However, as the array size increases, additional dividers and transmission lines are required, which can lead to increased losses.

7.3.5.2 Tree or Radial Feed Network

The tree feed is similar to the corporate feed, but it may have variable line lengths or unequal power division ratios. It is used when nonuniform amplitude stimulation is required to adjust the sidelobe level. This type of feed allows more flexible pattern shaping but is more challenging to design and match.

7.3.6 Mutual Coupling and Losses

In the majority of cases, mutual coupling is undesirable and affects the antenna performance. It reduces the gain and overall efficiency of the antenna. In antenna arrays, mutual coupling disturbs the current distribution and impacts the radiation shape and direction. The coupling in antenna arrays is mainly caused by improper spacing between the antenna structures. It also depends on the geometry and shape of individual elements. As a rule, if the spacing between adjacent elements is less than 0.5λ, coupling will be higher. Thus, unwanted back and side lobes will be generated and disturb the main pattern.

At high frequencies, such as mm-wave range, losses caused by the feeding lines become more prominent. These losses can be evaluated as:

$$L_{\text{total}} = L_{\text{dividers}} + L_{\text{lines}} + L_{\text{connectors}}. \tag{7.16}$$

The antenna performance in terms of total gain and efficiency can be enhanced by reducing these losses. Transmission line losses can be reduced by using shorter feed lines. Losses can be further minimized by using a low-loss substrate.

7.3.7 Applications of Parallel-fed Arrays

Parallel-fed arrays are widely used in applications where accurate beam control and high efficiency are required. Typical applications include the following:

- Phased array radar systems
- 5G base stations and mm-wave antennas
- Satellite communication and tracking systems
- Microwave imaging and sensing devices
- Automotive radar sensors

In mm-wave systems, parallel feeding is often realized using microstrip power dividers, SIW, or coplanar waveguide networks. The design must consider line width, dielectric constant, and fabrication tolerances to achieve good phase balance.

7.3.8 Summary

Parallel-fed arrays will be able to provide high flexibility and control of the radiation pattern. Designers can realize the same amplitude with different phase relationships between the elements, since all elements are fed in parallel. This technique will support a wide bandwidth, low sidelobe levels, and beam steering. Although the design is more complex and can suffer from higher losses, the performance advantages make it suitable for advanced mm-wave and high-frequency systems.

7.4 Combination-fed Arrays

7.4.1 Introduction

Combination-fed arrays, sometimes referred to as hybrid-fed arrays, are a unique kind of antenna array that use both parallel and series feeding methods. Combining the two approaches aims to maximize the advantages of each configuration while reducing its disadvantages. A combination-fed array has some parts that are connected in series and others that are fed in parallel. Good bandwidth, a steady amplitude distribution, and less feed loss are all balanced by this arrangement (Chi et al., 2017; Kothapudi and Kumar, 2021; Raj and Mandal, 2024).

The need for wideband, high-gain, and small antennas in contemporary wireless systems has made combination-fed arrays crucial. They are frequently utilized in satellite communications, radar sensors, and mm-wave communication. Depending on the required radiation pattern, the combination feeding structure can be constructed to generate either tapered amplitude or uniform excitation.

7.4.2 Principle of Combination Feeding

Combination feeding works on a simple but effective principle. The input power in a pure series-fed array passes sequentially through each element. This reduces feed complexity but restricts bandwidth and amplitude control. In a parallel-fed array, the power is split and sent to every element all at once, providing a large bandwidth but increasing losses owing to the feeding network.

In a combination-fed array, several elements are connected in series with the feed network, and the groups are fed in parallel, or vice versa. Initially, the power from the input port is divided into a few branches, with each branch feeding a small series chain of components. This mixed structure will provide better amplitude and phase control with less feed loss.

Mathematically, the total current at the nth element in a combination-fed array can be written as:

$$I_{mn} = A_m B_n e^{j(\phi_m + \psi_n)}, \tag{7.17}$$

where:

- A_m and ϕ_m represent the amplitude and phase contribution of the parallel-fed section.
- B_n and ψ_n represent the amplitude and phase of the series-fed subsection.

This expression shows that both the series and parallel parts contribute to the total excitation of the array.

7.4.3 Need for Combination-fed Arrays

Combination-fed arrays are used primarily to overcome some of the disadvantages of either of the individual feeding techniques. Although series-fed arrays are small, they offer limited beam control and narrow bandwidth. Parallel-fed arrays offer better performance but require intricate networks, which make fabrication more difficult and increase loss. A hybrid structure enables designers to achieve the following:

- Moderate feed complexity with improved amplitude control
- Wider bandwidth compared to a pure series feed
- Better efficiency compared to a full parallel network
- Enhanced beam steering capability
- Compact and low-profile structure

Therefore, combination-fed arrays provide an attractive balance between simplicity, performance, and efficiency.

7.4.4 Array Factor and Mathematical Model

The overall radiation pattern resulting from the combination of fields from each element is known as the array factor (AF). Since this is a combination-fed array, its array factor can be written as a double summation: one from the parallel part and one from the series part.

For an array with M parallel branches and N series elements in each branch:

$$\mathrm{AF}(\theta) = \sum_{m=1}^{M}\sum_{n=1}^{N} A_m B_n e^{j[k((m-1)d_p+(n-1)d_s)\sin\theta+\phi_m+\psi_n]}, \tag{7.18}$$

where:

- d_p is the spacing between the parallel branches.
- d_s is the spacing between elements in the series section.
- $k = 2\pi/\lambda$ is the wave number.

If the amplitude and phase are uniform for all elements, the equation simplifies to:

$$\mathrm{AF}(\theta) = AB\left[\frac{\sin\left(\frac{M}{2}(kd_p\sin\theta+\phi)\right)}{\sin\left(\frac{1}{2}(kd_p\sin\theta+\phi)\right)}\right]\left[\frac{\sin\left(\frac{N}{2}(kd_s\sin\theta+\psi)\right)}{\sin\left(\frac{1}{2}(kd_s\sin\theta+\psi)\right)}\right]. \tag{7.19}$$

This expression indicates that the overall array pattern is a product of two array factors: one for the parallel structure and one for the series structure.

7.4.5 Current and Phase Distribution

In a combination-fed array, the current distribution is more complex than in pure feeds. The amplitude of the current gradually decreases along the series chain due to power division, while it remains relatively uniform across parallel branches. The total excitation can be expressed as:

$$I_{mn} = I_0 \alpha^{n-1} e^{j(\beta_p(m-1)+\beta_s(n-1))} \tag{7.20}$$

where:

- α represents the current decay factor in the series-fed path.
- β_p and β_s represent the phase constants of the parallel and series feeds, respectively.

By adjusting α, β_p, and β_s, designers can shape the radiation pattern and control the beam direction. In practice, these parameters are controlled by the line lengths and impedance values in the feed network.

7.4.6 Typical Combination Structures

Combination-fed arrays can be built in several configurations depending on which feed type dominates.

7.4.6.1 Series-parallel Configuration

In this structure, small groups of elements are fed in parallel and connected in series. The advantages of this configuration are that the number of power dividers is lower than in the case of a fully parallel power divider, resulting in a more compact feed structure. The series component improves mechanical stability by reducing the number of feeding lines and connectors.

7.4.6.2 Parallel-series Configuration

This configuration represents the reverse of the previous one. In this configuration, the array is divided into a number of subarrays fed in parallel, with each subarray connected to the others in series. The method allows better control over phase delay and input impedance, which may be useful in the context of beam-steering applications.

7.4.6.3 Corporate–series Hybrid Feed

A corporate (binary tree) feed and a series section are combined in high-frequency designs, particularly at mm-wave bands. While the series section reduces network loss, the corporate part guarantees equal amplitude. SIW systems and patch array antennas frequently use this kind of feed.

7.4.7 Design Considerations and Equations

Designing a combination-fed array involves balancing amplitude, phase, impedance, and physical structure. Some key equations are mentioned as follows.

7.4.7.1 Power Division

If total input power is P_{in} and there are M parallel branches, each branch receives:

$$P_m = \frac{P_{\text{in}}}{M}. \tag{7.21}$$

For each series chain with N elements, the power to the nth element can be approximated as:

$$P_{mn} = P_m(1-\gamma)^{n-1}, \tag{7.22}$$

where γ represents the fraction of power radiated by each element.

7.4.7.2 Input Impedance

For a combination-fed structure, the input impedance is given by:

$$Z_{\text{in}} = \left(\sum_{m=1}^{M} \frac{1}{Z_{\text{branch},m}}\right)^{-1}. \tag{7.23}$$

Each branch impedance depends on the series chain, which can be expressed as:

$$Z_{\text{branch},m} = \sum_{n=1}^{N} Z_n e^{j\theta_n}. \tag{7.24}$$

Matching the network ensures minimal reflection and maximum power transfer.

7.4.8 Beam Steering and Pattern Control

One of the main advantages of combination-fed arrays is the ability to control both amplitude and phase more easily than with a single feeding method. Beam steering can be achieved by applying progressive phase shifts in either the series or parallel part.

If phase shift β_p is applied to the parallel section, the main beam angle θ_0 is given by:

$$kd_p \sin\theta_0 + \beta_p = 0. \tag{7.25}$$

Similarly, by adjusting the phase β_s in the series chain, the subarray beam can be steered to a different angle:

$$kd_s \sin\theta_1 + \beta_s = 0. \tag{7.26}$$

This double control mechanism gives flexible beam shaping, which is useful for adaptive or reconfigurable arrays.

7.4.9 Loss, Bandwidth, and Efficiency

Loss and bandwidth are always key performance parameters. In combination-fed arrays:

- **Feed loss** is lower than parallel-fed arrays because fewer dividers are used.
- **Bandwidth** is wider than series-fed arrays since impedance variation is smaller.
- **Efficiency** is high when using low-loss substrates and short line lengths.

The total feed loss can be estimated as:

$$L_{\text{total}} = L_{\text{divider}} + L_{\text{series\_line}} + L_{\text{connector}}. \tag{7.27}$$

Designers usually minimize the number of series elements per branch to keep losses small while maintaining compactness.

7.4.10 Mutual Coupling and Matching

Mutual coupling is another factor that affects array performance. In combination-fed arrays, the coupling is more complex since both parallel and series interactions occur. The coupling coefficient between two elements can be expressed as:

$$k_{ij} = \frac{E_i \cdot E_j^*}{|E_i||E_j|}. \tag{7.28}$$

To reduce coupling, the spacing between elements is usually kept around 0.5λ, and isolation structures such as defected ground or EBG patterns may be used.

Impedance matching is often done at two levels: within each series chain and at the input of the parallel branches. Quarter-wave transformers or tapered lines are common techniques to achieve wideband matching.

7.4.11 Comparison with Series and Parallel Arrays

A clear comparison helps to understand the role of combination feeding:

- Series-fed arrays are simple but have a narrow bandwidth.
- Parallel-fed arrays have a wide bandwidth but complex feeding networks.
- Combination-fed arrays achieve a compromise between both, offering good bandwidth, moderate complexity, and high gain.

Following is the comparison in table form (Table 7.1).

Table 7.1 Comparison of series-fed, parallel-fed, and combination-fed arrays.

Parameter	Series-fed	Parallel-fed	Combination-fed
Feed complexity	Low	High	Moderate
Bandwidth	Narrow	Wide	Moderate to wide
Amplitude control	Difficult	Easy	Flexible
Phase control	Limited	Excellent	Good
Loss in feed network	Low	High	Moderate
Fabrication difficulty	Simple	Complex	Moderate
Beam steering ability	Limited	Good	Very good
Typical application	Compact arrays	Phased arrays	Hybrid and mm-wave arrays

7.4.12 Applications

Combination-fed arrays are used in many modern systems that need compactness and high performance:

- 5G and 6G base station antennas
- Automotive and imaging radar sensors
- Satellite and aerospace communication systems
- Wearable and implantable mm-wave antennas
- Dual-polarized and reconfigurable array designs

In mm-wave design, hybrid feeding helps achieve low-loss feeding using SIW, coplanar waveguides, or microstrip power dividers combined with short series chains of patch antennas.

7.4.13 Summary

In summary, combination-fed arrays merge the benefits of both series and parallel feeding. They provide flexible amplitude and phase control, moderate loss, and improved bandwidth. The feeding network can be optimized to suit specific applications such as beam steering, high-gain arrays, or compact designs. The performance benefits make them perfect for contemporary communication systems, despite the fact that their structure is more complicated than that of simple feeds. In antenna engineering, the study of combination-fed arrays offers a crucial connection between conventional and cutting-edge feeding techniques.

7.5 Design Considerations for Antenna Arrays

Designing a single planar antenna is a straightforward task. However, an antenna involving several elements is not an easy task. It requires more attention by considering several factors. A well-designed array ensures that the system works as expected in real applications, such as 5G communication, radar, or satellite systems.

7.5.1 Element Spacing

The distance between elements in an array is very important. If the elements are too close, mutual coupling becomes strong, which can distort the radiation pattern. If they are too far apart, grating lobes may appear. In most cases, the element spacing d is chosen between 0.5λ and 0.8λ, where λ is the wavelength of operation.

For a linear array, the condition to avoid grating lobes is:

$$d \leq \frac{\lambda}{1 + |\sin\theta_0|}, \tag{7.29}$$

where θ_0 is the main beam angle. This ensures that the main lobe is dominant and unwanted lobes are minimized.

7.5.2 Amplitude Distribution

The amplitude of each element determines the strength of the radiated signal. A uniform amplitude distribution produces maximum directivity but higher side-lobe levels. Tapered amplitude distributions, such as Taylor or Dolph–Chebyshev, reduce side lobes but slightly decrease the main lobe strength.

The excitation amplitude A_n for a tapered array can be expressed as:

$$A_n = f(n, N), \tag{7.30}$$

where $f(n, N)$ is a function that depends on the element position n and the total number of elements N.

7.5.3 Phase Control

Phase control is essential for steering the beam in the desired direction. By introducing a progressive phase shift β between elements, the main lobe can be pointed at an angle θ_0. For a linear array:

$$\beta = -kd\sin\theta_0, \tag{7.31}$$

where $k = 2\pi/\lambda$ is the wave number. Proper phase control is crucial for phased arrays and beamforming systems.

7.5.4 Impedance Matching

Impedance matching ensures that maximum power is transferred from the source to the array elements. Mismatched impedance causes reflection and reduces efficiency. For arrays, the input impedance depends on the element type, spacing, and feed network.

Quarter-wave transformers are often used to match different impedances:

$$Z_t = \sqrt{Z_s Z_l}, \tag{7.32}$$

where Z_s is the source impedance, Z_l is the load impedance, and Z_t is the transformer impedance.

7.5.5 Mutual Coupling

Mutual coupling arises when elements are placed close to each other, affecting the current distribution and may distorting the radiation pattern. Designers mitigate coupling by maintaining appropriate element spacing, incorporating isolation structures, or selecting suitable element geometries.

The coupling coefficient between two elements can be approximated as:

$$k_{ij} = \frac{|V_i V_j^*|}{|V_i||V_j|}. \tag{7.33}$$

7.5.6 Feed Network Design

The feed network's design is crucial. Each element must receive the proper amplitude and phase from the network while losses are kept to a minimum. The selection of series, parallel, or combination feeds is determined by factors such as manufacturing restrictions, bandwidth, and efficiency. Attenuation is lessened by low-loss materials and thoughtful layout, particularly at high frequencies.

7.5.7 Bandwidth

The array's bandwidth is influenced by the element type, feed architecture, and mutual coupling. Communication systems are better off with a larger bandwidth.

7.5.8 Physical and Fabrication Constraints

Practical factors like cost, weight, and size of the antennas are crucial. Large arrays can be challenging to manufacture, and they tend to be bulky and heavy. Thus, effective antenna design necessitates a trade-off between functional requirements and overall performance. Compact designs are particularly crucial for wearable technology, satellites, and mm-wave systems.

7.6 Conclusion

This chapter presents the main concepts, structures, and practical design aspects of mm-wave planar antenna arrays. It explained why planar arrays are important at mm-wave frequencies, especially for applications that require high gain, directional radiation, beam steering, and easy integration with modern RF circuits. The chapter also discussed the working principles, strengths, and limitations of series-fed, parallel-fed, and combination-fed arrays, showing that each feeding method offers different benefits depending on the target application. In addition, important design factors such as element spacing, amplitude and phase control, impedance matching, mutual coupling, bandwidth, and fabrication constraints were described to highlight the trade-offs involved in array development. Overall, mm-wave planar arrays remain a key solution for future wireless systems, radar, sensing, and satellite platforms, and their successful design depends on achieving a careful balance between performance, complexity, efficiency, and practicality.

References

Jihoon Bang and Jaehoon Choi. A compact hemispherical beam-coverage phased array antenna unit for 5G mm-wave applications. *IEEE Access*, 8:139715–139726, 2020. https://doi.org/10.1109/ACCESS.2020.3013068.

Z. Chen and S. Otto. A taper optimization for pattern synthesis of microstrip series-fed patch array antennas. In *2009 European Wireless Technology Conference*, pages 160–163, 2009.

Taiyun Chi, Sensen Li, Jong Seok Park, and Hua Wang. A multifeed antenna for high-efficiency on-antenna power combining. *IEEE Transactions on Antennas and Propagation*, 65(12):6937–6951, 2017. https://doi.org/10.1109/TAP.2017.2764119.

Francesco Foglia Manzillo, Mauro Ettorre, Massimiliano Casaletti, Nicolas Capet, and Ronan Sauleau. Active impedance of infinite parallel-fed continuous transverse stub arrays. *IEEE Transactions on Antennas and Propagation*, 63(7):3291–3297, 2015. https://doi.org/10.1109/TAP.2015.2427874.

Sourav Ghosh, Gaurav Singh Baghel, and MV Swati. A low-profile, high gain, dual-port, planar array antenna for mm-wave powered 5G IoT systems. *AEU-International Journal of Electronics and Communications*, 155:154354, 2022.

J. Hirokawa and M. Ando. Efficiency of 76-GHz post-wall waveguide-fed parallel-plate slot arrays. *IEEE Transactions on Antennas and Propagation*, 48(11):1742–1745, 2000. https://doi.org/10.1109/8.900232.

J. Hirokawa, T. Miyagawa, M. Ando, and N. Goto. A waveguide-fed parallel plate slot array antenna. In *IEEE Antennas and Propagation Society International Symposium 1992 Digest*, vol. 3, pages 1471–1474, 1992. https://doi.org/10.1109/APS.1992.221611.

Yuefeng Hou, Yue Li, Zhijun Zhang, and Zhenghe Feng. A broadband and high-gain endfire antenna array fed by air-substrate parallel strip line. *IEEE Transactions on Antennas and Propagation*, 67(8):5717–5722, 2019. https://doi.org/10.1109/TAP.2019.2920332.

John Huang. A parallel-series-fed microstrip array with high efficiency and low cross-polarization. *Microwave and Optical Technology Letters*, 5(5):230–233, 1992.

Shuang Ji, Takashi Tomura, and Jiro Hirokawa. Fast analysis and bandwidth enhancement of the radiating subarray by method of moments for a parallel-plate slot array antenna with perpendicular-corporate feed. *IEEE Transactions on Antennas and Propagation*, 70(4):2645–2655, 2022. https://doi.org/10.1109/TAP.2021.3119059.

B. Jones, F. Chow, and A. Seeto. The synthesis of shaped patterns with series-fed microstrip patch arrays. *IEEE Transactions on Antennas and Propagation*, 30(6): 1206–1212, 1982. https://doi.org/10.1109/TAP.1982.1142963.

Mohsen Khalily, Rahim Tafazolli, T.A. Rahman, and M.R. Kamarudin. Design of phased arrays of series-fed patch antennas with reduced number of the controllers for 28-GHz mm-wave applications. *IEEE Antennas and Wireless Propagation Letters*, 15:1305–1308, 2016. https://doi.org/10.1109/LAWP.2015.

Venkata Kishore Kothapudi and Vijay Kumar. Hybrid-fed shared aperture antenna array for x/k-band airborne synthetic aperture radar applications. *IET Microwaves, Antennas & Propagation*, 15(1):93–102, 2021.

Chenfeng Li, Xiao-Wei Zhu, Pengfei Liu, Chao Yu, and Wei Hong. A metasurface-based multilayer wideband circularly polarized patch antenna array with a parallel feeding network for Q-band. *IEEE Antennas and Wireless Propagation Letters*, 18(6):1208–1212, 2019. https://doi.org/10.1109/LAWP.2019.2912624.

Mattia Maggi, Syrine Hidri, Loïc Marnat, Mauro Ettorre, Gerardo Orozco, Marc Margalef-Rovira, Christophe Gaquière, and Kamel Haddadi. Millimeter-wave phased arrays and over-the-air characterization for 5G and beyond: Overview on 5G mm-wave phased arrays and OTA characterization. *IEEE Microwave Magazine*, 23(5):67–83, 2022. https://doi.org/10.1109/MMM.2022.3148328.

Aqeel Hussain Naqvi and Sungjoon Lim. Review of recent phased arrays for millimeter-wave wireless communication. *Sensors*, 18(10):3194, 2018.

Daniele Pinchera. On the trade-off between the main parameters of planar antenna arrays. *Electronics*, 9(5):739, 2020.

D.M. Pozar and D.H. Schaubert. Comparison of three series fed microstrip array geometries. In *Proceedings of IEEE Antennas and Propagation Society International Symposium*, pages 728–731 vol.2, 1993. https://doi.org/10.1109/APS.1993.385244.

Arun Raj and Durbadal Mandal. Design and implementation of hybrid fed array antennae for sub-6 GHz and 5G mm-wave communication and wireless applications. *International Journal of Microwave and Wireless Technologies*, 16(4):605–624, 2024.

Harish Singh Rawat, Hema Singh, and RM Jha. Radar cross-section of a parallel-fed cylindrical array of dipoles. *Electromagnetics*, 36(5):287–304, 2016.

Bodhisatwa Sadhu, Xiaoxiong Gu, and Alberto Valdes-Garcia. The more (antennas), the merrier: A survey of silicon-based mm-wave phased arrays using multi-IC scaling. *IEEE Microwave Magazine*, 20(12):32–50, 2019. https://doi.org/10.1109/MMM.2019.2941632.

Feng Wei, Xinxin Liu, Xue-Zhi Ding, Xi-Bei Zhao, and Pei-Yuan Qin. A balanced filtering antenna array with high gain, steep selectivity, and multiradiation nulls parallel-fed by differential broadband network. *IEEE Transactions on Antennas and Propagation*, 71(12):9926–9931, 2023. https://doi.org/10.1109/TAP.2023.3311538.

Dixian Zhao, Peng Gu, Jiecheng Zhong, Na Peng, Mengru Yang, Yongran Yi, Jiajun Zhang, Pingyang He, Yuan Chai, Zhihui Chen, et al. Millimeter-wave integrated phased arrays. *IEEE Transactions on Circuits and Systems I: Regular Papers*, 68(10):3977–3990, 2021. https://doi.org/10.1109/TCSI.2021.3093093.

8

Millimeter-wave Reconfigurable Antennas Using FSS

8.1 Introduction

Today's wireless communication systems require high-speed transmission with low latency and fewer errors (Gunasekaran et al., 2025). Conventional microwave frequency band-based communication systems have several limitations. One of these limitations is the limited data rate. They cannot meet today's growing demands for fifth-generation (5G), satellite communication, and imaging systems. These applications need high data rates, which are not feasible with conventional microwave frequency band communications. The main reason behind this limitation is the limited available bandwidth at these frequency bands. High-data-rate communication is achievable at these frequency bands if more bandwidth becomes available. Due to the large available bandwidth at the millimeter-wave (mm-wave) frequency band, there is increasing interest in developing systems at these frequencies. Apart from several other good attributes of mm-wave technology, the radio frequency (RF) components and circuits, including antennas, are very compact in nature. This is because of the shorter wavelength associated with this frequency band. At the same time, the wide available bandwidth in this range supports extremely high data rates, which are crucial for future wireless standards such as 5G and sixth generation (6G).

In addition to their widespread use in wireless communication, mm-wave frequencies are also highly beneficial in radars, satellite links, security scanning, and biomedical imaging. Among many other areas, mm-wave radars are largely used in the automotive industry for driverless cars. Similarly, satellite communication in the mm-wave band can provide global coverage with faster data speeds. Imaging systems, such as airport security scanners or medical diagnostic equipment, also benefit from the fine resolution offered by short wavelengths. These examples highlight the broad importance of mm-wave systems in modern society.

8.1.1 Challenges of Antenna Design at mm-Wave Frequencies

While mm-wave technology offers many advantages, designing antennas for these frequencies is not a straightforward task and presents significant challenges. The miniaturization of antennas is both an opportunity and a challenge. It is well-known that mm-wave frequencies have a very small wavelength. This is because mm-wave frequencies are high (above 20–24 GHz). As the wavelength gets smaller, antennas become more compact in geometries. Because of the small size of antennas, their integration in compact devices becomes less challenging. At the same time, it raises other concerns, including high precision in fabrication. Even a small error in the antenna geometry can cause a large shift in operating frequency or a drop in performance.

Apart from requiring high precision in fabrication, it has several other issues. At mm-wave frequencies, high channel propagation loss and material loss are one of the big issues. Both of these have a direct relationship with the operating frequency. This is because free-space path loss increases significantly with frequency. As a result of free-space path loss, the coverage area becomes shorter. Similarly, conductive and dielectric losses in materials become more severe at mm-wave bands, lowering the antenna efficiency. This means designers must use advanced materials and accurate simulation tools to minimize these losses.

Fabricating an antenna at mm-wave is highly challenging, as the traces are typically fractions of a millimeter in volume. Antennas are generally fabricated using chemical etching, milling, or laser etching. A small misalignment between the layers or a small fabrication error highly impacts the antenna's resonance and matching. Special care should be taken to ensure that the prototype is developed without errors. In addition, transmission line losses become higher at these frequencies. Therefore, special attention is required for the feeding mechanism of the antenna as well. Moreover, the substrate losses become severe at these frequencies. A small variation in the substrate's dielectric constant or loss tangent deteriorates results or shifts resonance on either side. As a result, achieving consistent and reliable performance in large-scale production remains a challenge.

8.1.2 Why Reconfigurable Antennas

Conventionally, antennas have a fixed resonant frequency band where they operate. For example, Bluetooth antennas resonate at the fixed 2.4 GHz band with specific radiation properties. However, in real-world applications, the environment and system requirements fluctuate. For example, in a 5G network, the user device may need to switch between different frequency bands or adjust its beam direction to maintain a strong link. Like the above example, antennas for satellite communication need to change their radiation characteristics based on the intended mission. They should have the ability to change their radiation direction if required. Moreover, they should have adaptable polarization based on the needs of the mission.

Based on the examples given above, there is a strong need for reconfigurable antennas. Such antennas can adapt by altering their radiation pattern in a desired direction, changing their polarization, shifting their resonant frequency, adjusting their bandwidth, or performing a combination of these functions. An antenna that changes its polarization according to requirements is referred to as a polarization-reconfigurable antenna. Similarly, an antenna that adjusts its resonant frequency as needed is known as a frequency-reconfigurable antenna. Thus, reconfigurable antennas operate on-demand application requirements. They fulfill the requirement by changing their polarization, pattern, resonant frequency, or bandwidth. Because of their adaptability, they highly improve frequency usage and reduce the number of antenna elements. Thus, only one reconfigurable antenna can perform tasks of several fixed antennas. At high frequencies (mm-wave band), reconfigurable antennas become more vital because of high wireless link and directional beam requirements (Tanaka et al., 2025).

In addition, reconfigurable antennas also reduce the channel losses. This is because they have the ability to direct the radiation in the intended direction without needing extra effort. They can also switch between multiple bands to ensure connectivity in different scenarios. Thus, reconfigurable antennas are an essential building block for next-generation communication and sensing systems.

8.1.3 Role of Frequency Selective Surfaces

One of the promising techniques to enhance antenna performance and enable reconfigurability at mm-wave bands is the use of frequency selective surfaces (FSS) (Gong et al., 2025; Sun et al., 2025). An FSS is a periodic arrangement of metallic or dielectric elements, usually placed in a two-dimensional grid, as shown in Figure 8.1. Depending on the design, an FSS can act as a filter, allowing certain frequency bands to pass while blocking others (Hwang and Hong, 2025).

When integrated with antennas, FSS structures provide several benefits:

- FSS can be used as a superstrate to increase antenna gain and directivity, as shown in Figure 8.2.
- It can function as a reflector or absorber to control unwanted signals and reduce interference, as shown in Figure 8.3.
- With proper design, FSS can support polarization conversion and frequency tuning, as shown in Figure 8.4.

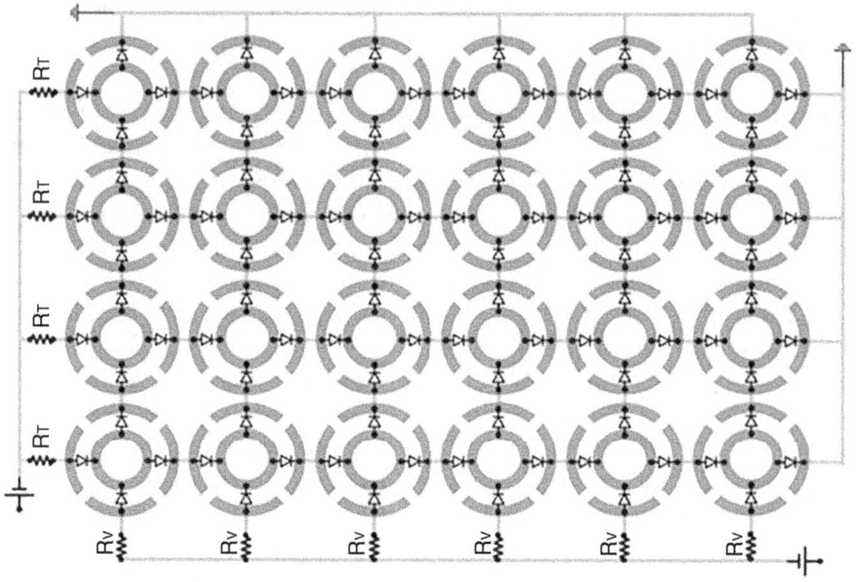

Figure 8.1 Reconfigurable FSS geometry. *Source*: Uslu et al. (2025)/MDPI/CC BY 4.0.

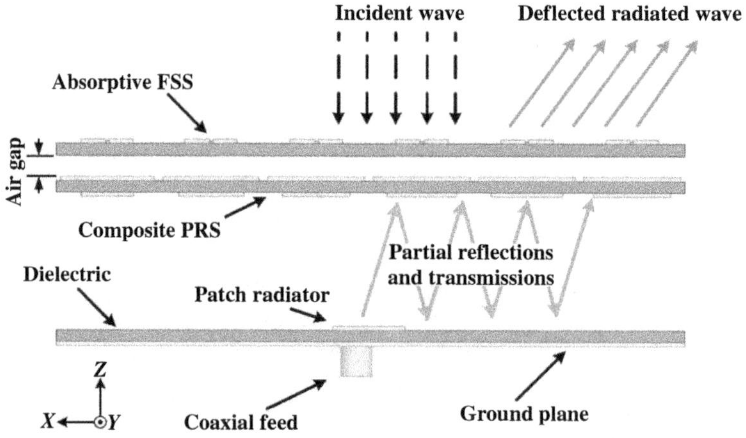

Figure 8.2 FSS as superstrate geometry. *Source*: Umair et al. (2021)/MDPI/CC BY 4.0.

The real power of FSS comes when it is combined with active components such as PIN diodes, micro-electro-mechanical systems (MEMS), liquid crystals, or graphene layers (Azemi et al., 2013). These active FSS designs enable dynamic control of transmission and reflection properties, which directly translates into reconfigurable antenna behavior (Ji et al., 2019). For instance, by switching the

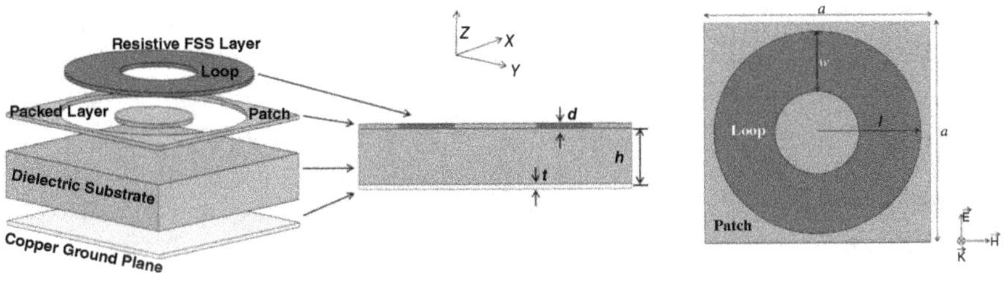

(a) (b)

Figure 8.3 (a) Geometry of unit cell of the absorber; (b) layout of the resonant structure of the absorber. *Source*: Lv et al. (2022)/MDPI/CC BY 4.0.

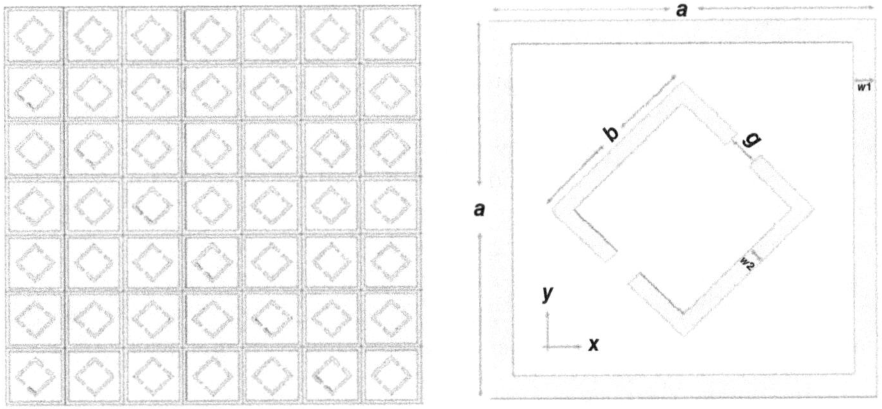

Figure 8.4 FSS 7×7 array and its unit cell. *Source*: Paramanik et al. (2025)/Springer Nature/CC BY 4.0.

state of the FSS elements, one can change the antenna's operating band, polarization, or beam direction without physically modifying the structure.

At mm-wave frequencies, FSS offers a lightweight, low-profile, and scalable solution. Because of their periodic nature, FSS structures can be designed to cover large surfaces, making them suitable for applications such as beam steering in 5G base stations, reconfigurable satellite reflectors, and smart surfaces for 6G networks.

8.1.4 Summary

In short, mm-wave technology is becoming the backbone of future wireless and imaging systems due to its ability to support high data rates and enable compact devices. However, antenna design at these frequencies faces challenges such as miniaturization, high losses, and strict fabrication requirements. Reconfigurable antennas provide an attractive solution by adapting their characteristics according to the scenario, improving performance and flexibility. FSS play a central role in this development by enhancing the gain, controlling the frequency response, and enabling reconfigurability through active elements. The combination of mm-wave technology, reconfigurable antennas, and FSS is therefore a key research direction for future communication, satellite, and imaging systems.

8.2 Fundamentals of FSS

8.2.1 Definition and Basic Principle of Operation

An FSS is a periodic structure consisting of metallic or dielectric patterns that manipulates the propagation or reflection of electromagnetic (EM) waves. In simple terms, an FSS functions similarly to an optical filter for visible light as a filter for radio waves. The FSS can block some frequency bands while permitting others to flow through, depending on how it is designed.

The FSS is all about the interaction of the EM waves with the FSS panel. In fact, the EM wave hits the FSS panel and induces current over there, which ultimately generates EM radiation. This induced radiation interferes with the incident wave, resulting in selective transmission or reflection at specific frequencies.

Mathematically, the periodicity of the FSS leads to a frequency-dependent transmission coefficient $T(\omega)$ and reflection coefficient $R(\omega)$, such that:

$$T(\omega) + R(\omega) + A(\omega) = 1, \tag{8.1}$$

where ω is the angular frequency and $A(\omega)$ represents the absorption loss. For an ideal lossless FSS, $A(\omega) = 0$, meaning that all the incident power is either transmitted or reflected.

8.2.2 Transmission vs. Reflection Mode FSS

FSS structures can be classified into two main categories:

- **Transmission Mode FSS:** In this case, the FSS allows certain frequency bands to pass through it while rejecting others. A typical example is an array of metallic patches on a dielectric sheet, which behaves like a bandpass filter. This type is widely used in radomes, satellite windows, and imaging systems where transparency at specific frequencies is required (Huang et al., 2022; Zhou et al., 2014). An example structure of transmissive/reflective FSS is shown in Figure 8.5 (Kumar et al., 2024).

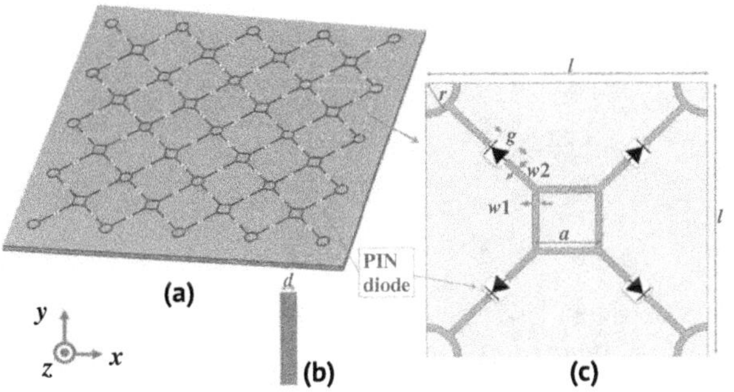

Figure 8.5 FSS geometry: (a) 3D view of the panel, (b) side view of the FSS panel, and (c) top view of the unit cell. *Source*: Kumar et al. (2024)/Springer Nature/CC BY 4.0.

- **Reflection Mode FSS:** Here, the FSS reflects certain frequency bands while allowing others to pass. Slot-type unit cells, for example, often behave in this manner. Reflection-mode FSS can be used as a frequency-selective reflector in antenna systems to enhance the gain or suppress unwanted bands (Ji et al., 2019; Yan et al., 2019).

The choice between transmission and reflection mode depends on the application. For instance, in a reconfigurable antenna, the reflection-mode FSS may be used as a back surface to direct the radiation, while in communication windows, the transmission-mode FSS ensures selective transparency.

8.2.3 Equivalent Circuit Models for FSS

To simplify analysis, FSS unit cells are often represented using equivalent LC circuit models. The geometry of the metallic element determines whether the response is capacitive, inductive, or a combination of both.

- **Patch-type FSS** acts like a capacitive grid. The small gaps between adjacent patches introduce capacitance, which dominates the behavior (Foroozesh and Shafai, 2010). The equivalent circuit can be modeled as a capacitor in series with the transmission line. The circuit model of such a configuration is shown in Figure 8.6.
- **Slot-type FSS** acts as an inductive grid. The slots allow current to flow around them, producing an inductive effect (Liang et al., 2015). The equivalent circuit can be modeled as an inductor in series with the transmission line.

A simple LC resonator model can describe the resonance of the FSS:

$$f_0 = \frac{1}{2\pi\sqrt{LC}}, \tag{8.2}$$

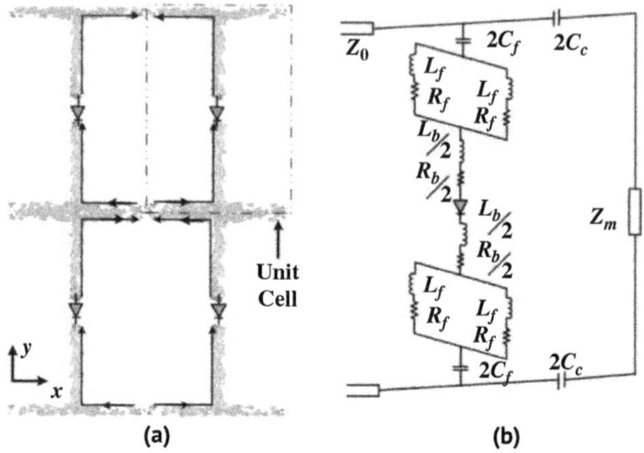

Figure 8.6 (a) Surface current distribution; (b) equivalent circuit of the active FSS. *Source*: Zhao et al. (2019)/IEEE/CC BY 4.0.

where f_0 is the resonance frequency, L is the effective inductance, and C is the effective capacitance of the unit cell. By adjusting the geometry of the FSS element, one can tune L and C, thus shifting the resonance frequency.

This circuit analogy helps designers to understand how changing parameters such as element size, gap width, or substrate thickness influence the frequency response.

8.2.4 Types of Unit Cells

The performance of an FSS is determined by the geometry of its repeating unit cell. Several common types are widely used:

- **Patch Elements**: Square, circular, or rectangular patches arranged in a periodic array. These generally produce a bandstop response.
- **Slot Elements**: Periodic slots cut into a conductive sheet. These typically result in a bandpass response.
- **Dipole Elements**: Arrays of dipole-like structures that provide frequency-selective behavior similar to wire grids (Garcia-Vigueras et al., 2010; Maci et al., 2005). This type of structure is shown in Figure 8.7.
- **Jerusalem Cross**: A cross-shaped element with additional arms, offering dual-polarization selectivity and multiband operation (Hosseini and Hakkak, 2008; Hosseinipanah and Wu, 2008).
- **Fractal-based Unit Cells**: Use self-similar geometries (e.g., Koch curve, Hilbert curve) to achieve multiband and wideband responses (Romeu and Rahmat-Samii, 1999; Solunke and Kothari, 2025). This type of structure is shown in Figure 8.8.

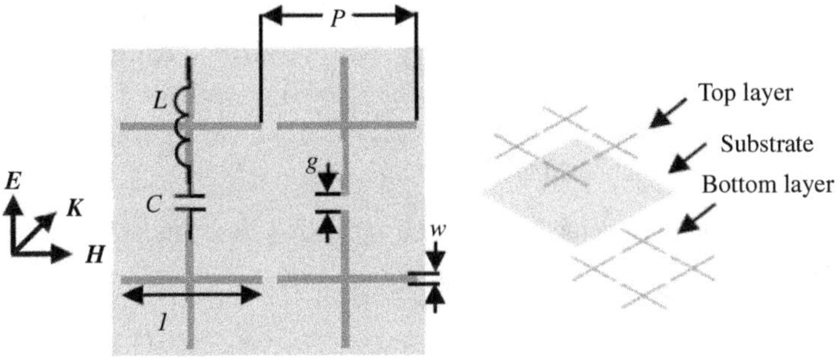

Figure 8.7 Dipole-based FSS. *Source*: Zouaoui et al. (2021)/IEEE/CC BY 4.0.

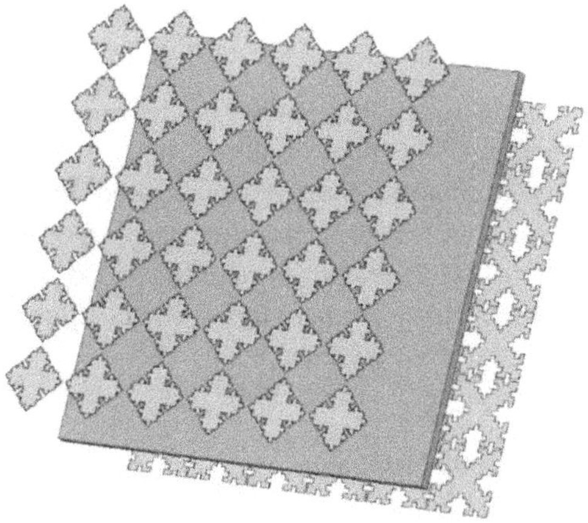

Figure 8.8 Geometry of the fractal FSS. *Source*: Pan and Dong (2023)/MDPI/CC BY 4.0.

The choice of unit cell depends on the desired filter characteristics. For example, patch arrays are often used in radomes, while fractal-based unit cells are suitable for wideband reconfigurable antennas.

8.2.5 Bandpass, Bandstop, and Polarization-dependent Designs

- **Bandpass FSS**: Designed to allow a certain band of frequencies to pass while blocking all others. Slot arrays are commonly used to realize bandpass behavior.
- **Bandstop FSS**: These reject a specific frequency band while passing others. Patch arrays are a typical example. Bandstop FSS are useful for filtering out unwanted signals in antenna systems.

- **Polarization-dependent FSS**: Some FSS designs behave differently depending on the polarization of the incident wave. For instance, linear dipole FSS may pass horizontally polarized waves while blocking vertically polarized ones. Polarization-selective FSS are important in satellite and radar systems, where cross-polarization control is required.

Mathematically, polarization selectivity can be expressed in terms of the transmission coefficients for orthogonal polarizations:

$$T_x(\omega) \neq T_y(\omega), \tag{8.3}$$

where $T_x(\omega)$ and $T_y(\omega)$ are the transmission responses for x- and y-polarized waves, respectively.

8.2.6 Summary

FSS provide a powerful and flexible way to control EM waves. By using periodic unit cells such as patches, slots, or fractals, FSS can be designed to act as bandpass or bandstop filters, or even polarization-sensitive surfaces. The equivalent LC circuit models give a simple way to predict resonance frequencies and design parameters. Depending on the application, an FSS may operate in transmission or reflection mode. These fundamental concepts form the base for advanced applications such as reconfigurable antennas at mm-wave frequencies.

8.3 mm-Wave Reconfigurable Antennas

8.3.1 Overview of Reconfigurability

Antennas are one of the most important parts of any wireless system. Traditional antennas are usually designed to work at a fixed frequency and with fixed characteristics. However, in modern communication systems, the requirements change with time and environment. For example, a mobile phone may need to connect to different frequency bands, or a satellite antenna may need to change its polarization depending on the mission. To meet these changing needs, researchers have developed reconfigurable antennas.

A reconfigurable antenna is able to change its properties dynamically. This change can happen in frequency, polarization, radiation pattern, or even gain. At mm-wave bands, reconfigurable antennas are very attractive because they can help reduce losses, improve link quality, and provide higher flexibility in systems such as 5G, 6G, radar, and imaging devices.

8.3.2 Frequency Reconfigurability

Frequency reconfigurability means that the antenna can change its operating band. For instance, the same antenna can be tuned to work at 28 GHz for 5G applications and then shifted to 38 GHz for backhaul links. This type of flexibility avoids the need for multiple separate antennas.

The principle is based on adjusting the electrical length of the antenna or changing the resonant conditions. When an antenna resonates, the input reactance becomes zero, and the condition can be written as:

$$X_{\text{in}}(f) = 0. \tag{8.4}$$

By modifying the structure using switches or tunable materials, the resonance frequency f can be shifted. Frequency reconfigurable antennas are very useful for cognitive radios, adaptive satellite communication, and dynamic spectrum allocation in 5G/6G networks.

8.3.3 Polarization Reconfigurability

Polarization refers to the orientation of the electric field in an EM wave. Some systems require linear polarization, while others need circular polarization. A polarization-reconfigurable antenna can switch between these states depending on the application.

For example, in satellite communication, right-hand circular polarization (RHCP) is often used for uplink, while left-hand circular polarization (LHCP) may be used for downlink. A reconfigurable antenna can provide both without needing two separate designs.

The ability to control polarization also improves the system's reliability. If the channel suffers from polarization mismatch, the antenna can adjust itself to the correct orientation, ensuring a stronger signal.

8.3.4 Radiation Pattern Reconfigurability

The radiation pattern of an antenna shows how it radiates power in space. A fixed antenna always radiates in the same way, but a reconfigurable antenna can change its pattern.

For example, an antenna can be designed to radiate in a broadside direction at one time and then switch to endfire at another time. This flexibility is especially important in mm-wave communication, where the line-of-sight path can be blocked easily by walls or obstacles. By changing its pattern, the antenna can redirect energy to a clearer path.

Mathematically, the radiation pattern is represented as $F(\theta, \phi)$, where θ and ϕ are angular coordinates. Reconfigurability allows modification of $F(\theta, \phi)$ to optimize coverage and link quality.

8.3.5 Gain and Beam Steering Reconfigurability

At mm-wave bands, the propagation loss is significantly high. This means antennas need a higher gain to maintain a strong link. A reconfigurable antenna can not only change its gain but also steer the main beam in the desired direction.

Beam steering is especially important for 5G base stations and vehicular communication systems. Instead of radiating energy in all directions, the antenna focuses the beam toward the user, which increases signal strength and reduces interference.

In phased array systems, beam steering is achieved by adjusting the relative phase of each element. The steering angle θ_0 can be controlled using:

$$\theta_0 = \sin^{-1}\left(\frac{\lambda \Delta\phi}{2\pi d}\right), \tag{8.5}$$

where $\Delta\phi$ is the progressive phase shift, λ is the wavelength, and d is the spacing between elements.

Reconfigurable FSS-based antennas can also achieve beam steering by altering the reflection or transmission properties of the surface.

8.3.6 Techniques for Reconfiguration

A variety of techniques are employed to realize reconfigurability in antennas, with a particular focus on mm-wave bands:

- **PIN Diodes:** Simple semiconductor switches that can turn parts of the antenna on or off. They are widely used because of their low cost and ease of integration, but they introduce losses at mm-wave frequencies. The PIN diode-based FSS structure is shown in Figure 8.9.
- **Varactor Diodes:** These are variable capacitors controlled by a bias voltage. By tuning the capacitance, the resonance frequency of the antenna can be changed. At mm-wave, however, parasitics of varactors can limit the system performance (Mias and Tsakonas, 2005). The varactor diode-based FSS structure is shown in Figure 8.10.
- **MEMS:** MEMS switches use tiny movable structures to connect or disconnect parts of the antenna. They have low insertion loss and high isolation at mm-wave, making them very suitable (Schoenlinner et al., 2004).

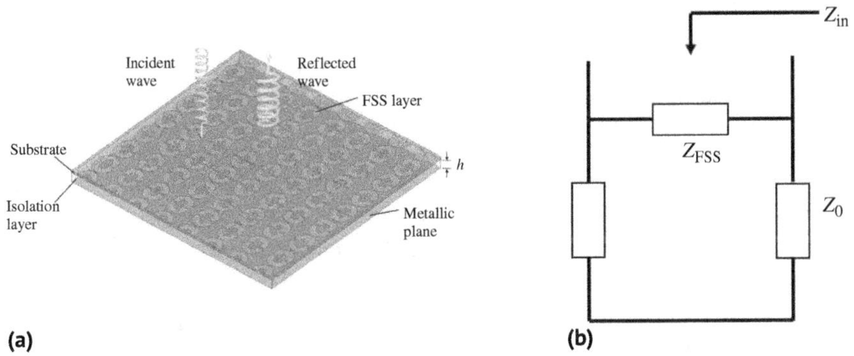

Figure 8.9 (a) The geometry of tunable FSS; (b) equivalent circuit model. *Source*: Lin and Fang (2025)/MDPI/CC BY 4.0.

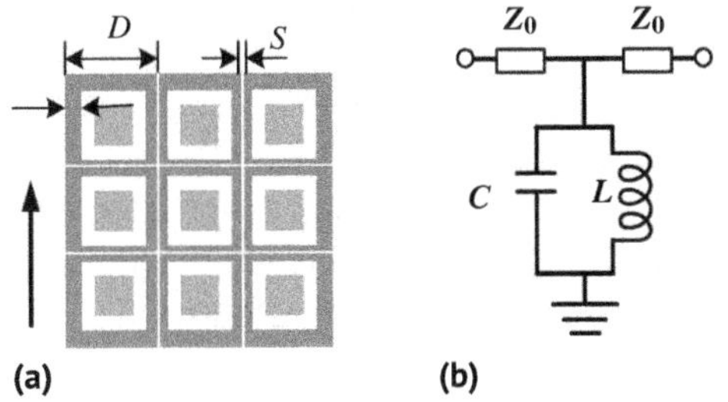

Figure 8.10 (a) The geometry of FSS; (b) circuit model. *Source*: Gong et al. (2025)/MDPI/CC BY 4.0.

- **Liquid Crystals:** These materials change their dielectric properties when an electric field is applied. They can be used for tunable lenses and surfaces, though their response time is relatively slow (Wu et al., 2020).
- **Graphene:** A two-dimensional material with tunable conductivity. By applying a bias voltage, the surface impedance of graphene can be controlled, making it highly attractive for reconfigurable antennas in mm-wave and terahertz (THz) bands (Chen et al., 2020).
- **Phase-change Materials (e.g., VO_2):** These materials can switch between insulating and metallic states with thermal or electrical stimulus. They are promising for reconfigurable reflectarrays and FSS designs (Long et al., 2020).

Each technique has advantages and limitations. The choice depends on the required reconfiguration speed, frequency range, and fabrication technology.

8.3.7 Challenges at mm-Wave Bands

Although reconfigurable antennas offer great potential, integrating them at mm-wave frequencies is not easy. Some of the main challenges include:

- **Biasing Networks:** Reconfigurable elements such as diodes or MEMS require bias lines to operate. At mm-wave bands, adding biasing networks without disturbing the antenna performance is very challenging. Bias lines can introduce unwanted parasitic radiation and affect impedance matching.
- **Parasitics:** At high frequencies, even small parasitic capacitances and inductances have a big impact. Components such as diodes and solder joints introduce unwanted reactances that shift the operating frequency.
- **Fabrication Limits:** The small wavelength at mm-wave requires extremely precise fabrication. A slight error in etching or assembly can cause large performance degradation. Advanced techniques such as photolithography and micromachining are often required, though they increase cost.
- **Losses:** Active components such as PIN diodes and varactors introduce insertion loss. At mm-wave, this loss is more significant due to the shorter wavelength and higher surface resistance of conductors.
- **Reliability:** Materials like liquid crystals or MEMS may suffer from slow response or mechanical wear. Ensuring long-term reliability in real environments (temperature changes, vibrations, humidity) is still a concern.

8.3.8 Summary

Reconfigurable antennas at mm-wave bands represent an important step toward flexible and adaptive wireless systems. They can modify their frequency, polarization, radiation pattern, and gain to suit different environments and applications. Techniques such as PIN diodes, MEMS, liquid crystals, graphene, and phase-change materials are used to achieve reconfigurability, though each comes with challenges. Biasing networks, parasitics, fabrication tolerances, and losses make mm-wave reconfigurable antenna design a more complex task.

Despite these challenges, research in this field is growing rapidly, driven by the needs of 5G, 6G, satellite communication, radar, and imaging systems. The next section will discuss how FSS can be integrated with these antennas to overcome limitations and provide advanced reconfigurable solutions.

8.4 Integration of FSS with mm-Wave Reconfigurable Antennas

The integration of FSS with mm-wave reconfigurable antennas has emerged as a key area of research in modern wireless communication systems. FSS structures, when combined with antennas, can improve the gain, control beam patterns, allow frequency switching, and support polarization reconfigurability. At mm-wave frequencies, where antennas face issues like higher losses, limited efficiency, and design complexity, FSS provides a flexible solution to overcome many of these limitations. In this section, the different roles of FSS in mm-wave reconfigurable antennas are thoroughly discussed, with the inclusion of mathematical models where relevant.

8.4.1 FSS as Superstrate for Gain Enhancement

One of the most popular uses of FSS is as a superstrate layer placed above an antenna. The idea is similar to using a lens. The FSS modifies the wavefronts radiated by the antenna, leading to higher directivity and gain.

When an antenna radiates, the electric field at a distance r can be written as:

$$E(\theta, \phi) = \frac{I_0 l e^{-jkr}}{4\pi r} \cos\theta, \tag{8.6}$$

where I_0 is the current, l is the length of the radiator, $k = 2\pi/\lambda$ is the wave number, and θ is the observation angle.

By introducing an FSS superstrate, the phase of $E(\theta, \phi)$ is altered in such a way that more radiation is directed along the broadside direction. This improves the directivity D, which is given as:

$$D = \frac{4\pi U(\theta, \phi)}{P_{\text{rad}}}, \tag{8.7}$$

where $U(\theta, \phi)$ is the radiation intensity and P_{rad} is the total radiated power.

The distance between the antenna and the FSS is often optimized at $\lambda/4$ or $\lambda/2$ to maximize constructive interference and achieve a higher gain.

8.4.2 FSS as a Reconfigurable Reflector/Absorber for Beam Control

An FSS reflector works by controlling the reflection coefficient Γ, defined as:

$$\Gamma = \frac{Z_{\text{FSS}} - Z_0}{Z_{\text{FSS}} + Z_0}, \tag{8.8}$$

where Z_{FSS} is the surface impedance of the FSS and Z_0 is the free-space impedance (377 Ω).

By tuning Z_{FSS} using active elements (e.g., PIN diodes, varactors), the phase of Γ changes, and this directly shifts the beam direction of the antenna. Similarly, for an absorber-type FSS, the condition is:

$$Z_{\text{FSS}} \approx Z_0, \tag{8.9}$$

which minimizes reflection and maximizes absorption.

This property is useful in reducing interference and controlling unwanted radiation in mm-wave systems.

8.4.3 FSS-based Polarization-reconfigurable Antennas

Polarization control can be achieved by designing anisotropic FSS unit cells. A polarization conversion ratio (PCR) can be defined as:

$$\mathrm{PCR} = \frac{|E_{\text{cross}}|^2}{|E_{\text{co}}|^2 + |E_{\text{cross}}|^2}, \tag{8.10}$$

where E_{co} is the co-polarized field and E_{cross} is the cross-polarized field.

By tuning the FSS structure, the antenna can switch between linear polarization (LP), RHCP, and LHCP. This is especially important for adaptive communication links at mm-wave bands.

8.4.4 FSS Metasurfaces for Multiband and Wideband Operation

Multiband operation of FSS is achieved by designing unit cells that resonate at multiple frequencies. The resonance condition can be approximated as:

$$f_r = \frac{c}{2L\sqrt{\epsilon_{\text{eff}}}}, \tag{8.11}$$

where c is the speed of light, L is the effective length of the unit cell, and ϵ_{eff} is the effective dielectric constant.

For wideband operation, stacked or fractal FSS designs are used to create multiple overlapping resonances, leading to a broader passband.

This ensures stable antenna performance across wide frequency ranges, which is required for 5G/6G and high-data-rate mm-wave applications.

8.4.5 Hybrid Integration: FSS with Phased Arrays, RIS, and MIMO Antennas

8.4.5.1 FSS with Phased Arrays

In phased arrays, beam steering is achieved using progressive phase shifts. For an N-element array, the array factor is:

$$\mathrm{AF}(\theta) = \sum_{n=1}^{N} e^{j(n-1)(kd\cos\theta+\beta)}, \tag{8.12}$$

where d is the element spacing and β is the progressive phase shift.

Integrating an FSS with phased arrays enhances gain and reduces side lobes by acting as a filtering layer.

8.4.5.2 FSS with RIS

Reconfigurable intelligent surfaces (RIS) are large-scale intelligent metasurfaces often built using tunable FSS elements. The reflection response of RIS can be modeled as:

$$E_{\text{out}} = \sum_{m=1}^{M} A_m e^{j\phi_m}, \tag{8.13}$$

where A_m and ϕ_m are the amplitude and phase response of the mth unit cell.

By programming ϕ_m, RIS can dynamically control wireless channels, improving link quality in 6G mm-wave networks.

8.4.5.3 FSS with MIMO Antennas

For multiple-input multiple-output (MIMO) antennas, isolation between ports is crucial. The isolation can be quantified using the transmission coefficient:

$$S_{21} = \frac{P_{\text{transmitted}}}{P_{\text{incident}}}. \tag{8.14}$$

An FSS layer between antenna elements helps reduce mutual coupling, thereby improving S_{21} performance. This increases channel capacity in mm-wave MIMO systems.

8.4.6 Conclusion

The integration of FSS with mm-wave reconfigurable antennas provides multiple benefits: improved gain, beam steering, polarization control, and wideband/multiband operation. The mathematical models of reflection, transmission, and polarization conversion explain how FSS impacts antenna performance. Furthermore, hybrid integration with phased arrays, RIS, and MIMO makes FSS highly relevant for next-generation mm-wave wireless systems. While fabrication and integration challenges remain, the advantages of flexibility and high performance make FSS a promising technology for the future.

8.5 Design Methodologies

The design of FSS for mm-wave reconfigurable antennas requires careful planning and a step-by-step methodology. Since mm-wave systems operate at very high frequencies, even a small variation in geometry or material can cause large changes in performance. This section explains the design methods in detail, covering unit cell design, simulation techniques, parametric analysis, biasing of active components, and fabrication challenges. The discussion is presented in a simple and practical way to give the reader a clear idea of how FSS structures are designed for real-world applications.

8.5.1 Unit Cell Design and Periodicity Considerations

The first step in designing an FSS is to select the basic unit cell. The unit cell is the smallest repeating pattern that forms the overall surface. At mm-wave frequencies, the size of the unit cell becomes very small, often in the range of a few hundred micrometers to a few millimeters. The geometry of the unit cell decides whether the FSS will behave like a bandpass filter, bandstop filter, or a polarization converter.

Common unit cell shapes include square patches, circular patches, slots, dipoles, Jerusalem crosses, and fractal geometries. Each shape has its own advantages. For example, patch-based unit cells usually provide bandstop characteristics, while slot-based designs often create bandpass responses. Fractal shapes can be used for multiband operation.

The periodicity of the unit cell, which is the spacing between adjacent cells, is another important factor. The periodicity p is usually kept less than half the wavelength ($p < \lambda/2$) at the design frequency to avoid grating lobes. If the periodicity is too large, unwanted diffraction modes will appear, which reduces the efficiency and stability of the FSS.

The resonance frequency of the unit cell can be approximated by:

$$f_r = \frac{c}{2L\sqrt{\epsilon_{\text{eff}}}}, \tag{8.15}$$

where L is the effective length of the resonant structure, c is the speed of light, and ϵ_{eff} is the effective dielectric constant of the substrate. This simple formula helps designers estimate the starting dimensions before fine-tuning in simulations.

8.5.2 Simulation Approaches: CST, HFSS, and Equivalent Circuit Models

Simulation is a very important part of designing FSS for mm-wave antennas. Since physical fabrication and testing at mm-wave frequencies are costly and time-consuming, most of the optimization is performed through EM simulators.

Popular commercial software includes computer simulation technology (CST) Microwave Studio and Ansys high-frequency structure simulator (HFSS). These tools allow full-wave simulations based on the finite integration technique (FIT) or finite element method (FEM). In these simulations, a single unit cell is placed inside a periodic boundary condition (PBC), and a plane wave is used as an excitation. This setup mimics an infinite array of unit cells, which is a good approximation for practical FSS layers.

Equivalent circuit models are also used in the early design stages. For example, a patch-type unit cell can be modeled as an LC circuit:

$$Z_{\text{eq}} = j\omega L + \frac{1}{j\omega C}, \tag{8.16}$$

where L represents the inductive response due to current loops and C represents the capacitive gap between adjacent elements. Such simple models help in quickly predicting resonance frequencies and bandwidth before running heavy simulations.

CST and HFSS provide detailed information such as reflection coefficient (S_{11}), transmission coefficient (S_{21}), surface current distribution, and phase response. This information is then used to refine the geometry of the unit cell.

8.5.3 Parametric Analysis of Unit Cell Geometry and Materials

Once the initial design is set, a parametric analysis is performed to understand the effect of different parameters on performance. The most common parameters include:

- **Unit Cell Geometry:** The length, width, and shape of the resonator directly affect the resonance frequency and bandwidth. For example, increasing the length of a patch lowers the resonance frequency, while reducing the gap between elements increases capacitance and shifts the frequency.
- **Dielectric Substrate:** The type of substrate used is critical at mm-wave. Materials with low loss tangent (e.g., Rogers RO3003, quartz) are often chosen. The thickness of the substrate also changes the coupling between the FSS and the antenna.
- **Periodic Spacing:** Adjusting the periodicity controls diffraction effects and bandwidth. A smaller periodicity improves angular stability, which is important at mm-wave frequencies.
- **Metal Thickness:** At mm-wave frequencies, the skin depth of the conductor becomes very small, often a few micrometers. Therefore, the metal layer must be thick enough to reduce ohmic losses.

During the parametric analysis, one parameter is varied while keeping others constant. The response is observed in terms of S_{11} and S_{21}, with an optimized design achieved after several iterations.

8.5.4 Biasing Techniques for Active FSS Elements

To achieve reconfigurability, active components such as PIN diodes, varactors, MEMS, or graphene are embedded into the FSS unit cells. These elements require biasing networks, which provide DC voltage or current to control the state of the device. Designing biasing networks at mm-wave frequencies is challenging, because the wires and vias used for biasing can disturb the RF performance of the FSS.

One common solution is to use high-impedance bias lines that do not interfere much with the RF signals. Another approach is to embed bias lines inside the substrate using multilayer technology. In some designs, optical biasing is used, where the active material is controlled by light instead of electrical wires, which removes the need for physical bias lines.

At mm-wave frequencies, biasing networks must be very low-profile. If the bias network is too large, it can act as an unwanted resonator or scatterer. Therefore, compact designs with minimal parasitics are preferred. For MEMS-based FSS, electrostatic actuation is often used, which requires very low power but provides reliable switching.

8.5.5 Fabrication Challenges: Lithography, 3D Printing, and Flexible Substrates

Fabricating FSS for mm-wave applications is not straightforward. The small size of the unit cells requires high precision, and even small fabrication errors can cause significant frequency shifts. The main fabrication methods include the following.

8.5.5.1 Lithography

Photolithography is the most common method for fabricating FSS at mm-wave frequencies. It provides high resolution and allows features in the micrometer range. However, lithography is costly and requires cleanroom facilities. Any misalignment or undercut during the process can reduce the performance of the final device.

8.5.5.2 3D Printing

With recent advances, 3D printing has become an attractive option for FSS fabrication. It allows rapid prototyping and low-cost production, especially for complex or multilayer structures. However, the resolution of 3D printing is still lower compared to lithography, and the surface roughness can cause additional losses at mm-wave.

8.5.5.3 Flexible Substrates

Flexible and conformal FSS are gaining attention for wearable and implantable applications. Substrates like polyimide or PET can be used to make bendable FSS. The challenge here is that bending changes the periodicity and unit cell geometry, which shifts the resonance. Therefore, designs must consider deformation effects from the start.

Another fabrication challenge at mm-wave is the alignment of multiple layers. In multilayer FSS structures, even a small misalignment of a few micrometers can degrade performance. Thermal stability is also important, because mm-wave systems often operate in harsh environments.

8.5.6 Summary

The design of FSS for mm-wave reconfigurable antennas requires careful planning of unit cell geometry, substrate selection, and periodicity. Simulation tools such as CST and HFSS are essential for accurate design, while equivalent circuit models help in initial predictions. A parametric analysis allows fine-tuning of performance, and active biasing techniques provide reconfigurability. Finally, fabrication methods such as lithography, 3D printing, and flexible substrates each have their own advantages and challenges. By following these design methodologies, high-performance FSS structures can be created to support advanced mm-wave communication systems.

8.6 Performance Metrics

When we design mm-wave reconfigurable antennas with FSS, we must check their performance through many measurable values. These values, called performance metrics, help us to understand how good the antenna works in terms of efficiency, reliability, and cost. In this section, the important metrics are explained, with a focus on reflection and transmission coefficients, bandwidth, gain, efficiency, and isolation. Trade-offs between complexity, size, and cost are also discussed.

8.6.1 Reflection and Transmission Coefficient (S-parameters)

The most basic metric for antennas and FSS is the S-parameters. At mm-wave frequencies, the common parameters are S_{11} (reflection coefficient) and S_{21} (transmission coefficient).

The reflection coefficient S_{11} tells us how much power is reflected back from the antenna input. If S_{11} is low (usually below −10 dB), it means most of the power is accepted by the antenna. Mathematically, it can be written as:

$$\Gamma = \frac{Z_{\text{in}} - Z_0}{Z_{\text{in}} + Z_0}, \tag{8.17}$$

where Z_{in} is the input impedance of the antenna and Z_0 is the characteristic impedance of the feed line, usually 50 Ω.

The transmission coefficient S_{21} shows how much signal passes through the FSS. For example, in bandpass FSS, we want high S_{21} at the operating frequency. In bandstop FSS, we want low S_{21} at unwanted frequencies.

In practice, S-parameters at mm-wave are affected by fabrication errors, surface roughness, and material losses. Thus, achieving a clean and deep S_{11} dip at mm-wave bands is a big challenge.

8.6.2 Bandwidth and Frequency Agility

Bandwidth is another key metric. It shows the frequency range over which the antenna or FSS can operate with acceptable performance. At mm-wave, a wide bandwidth is important for 5G and 6G systems to support high data rates.

The fractional bandwidth (FBW) is often calculated as:

$$\text{FBW} = \frac{f_H - f_L}{f_c} \times 100\%, \tag{8.18}$$

where f_H is the upper frequency, f_L is the lower frequency, and f_c is the center frequency.

Reconfigurable antennas and FSS can change their operating frequency using active components like PIN diodes, varactors, MEMS, or graphene. This ability is called frequency agility. It means the system can cover multiple bands without changing the physical size.

However, at mm-wave, the range of frequency tuning is sometimes limited because even small changes in geometry or bias voltage can shift the resonant frequency by a large amount. Also, parasitic effects of diodes and bias networks may reduce the bandwidth.

8.6.3 Gain Enhancement and Beam Shaping

The gain is a measure of how much power the antenna can radiate in a specific direction compared to an isotropic radiator. At mm-wave frequencies, antennas are usually small, so the natural gain is low.

FSS can help to enhance the gain by acting as a superstrate or reflector. For example, when an FSS is placed above an antenna, it can reflect the waves in phase and create constructive interference, which increases the antenna gain.

The gain G can be written as:

$$G(\theta, \phi) = \eta \cdot D(\theta, \phi), \tag{8.19}$$

where η is the efficiency and D is the directivity.

Beam shaping is another important metric. It means the antenna can control the direction and shape of its radiation beam. With reconfigurable FSS, the beam can be steered toward a user or scanned across space. This is especially useful in phased arrays and 5G base stations.

Challenges include maintaining stable gain across different frequencies and ensuring that beam scanning does not introduce high side-lobe levels.

8.6.4 Efficiency and Losses at mm-Wave

Efficiency is one of the most critical metrics at mm-wave. It tells us how much of the input power is converted into radiated power. Losses can happen due to conductor resistance, dielectric losses, surface roughness, and parasitic effects of reconfigurable components.

The radiation efficiency η_{rad} is given as:

$$\eta_{\text{rad}} = \frac{P_{\text{rad}}}{P_{\text{in}}}, \tag{8.20}$$

where P_{rad} is the radiated power and P_{in} is the input power.

At mm-wave, efficiency is often reduced because metals like copper show higher surface resistance at high frequencies. Also, active devices like PIN diodes and MEMS introduce insertion loss. To improve efficiency, designers use low-loss substrates, high-quality metals, and optimized biasing networks.

A balance must be kept: adding more reconfigurable features usually lowers the efficiency, but it also increases functionality.

8.6.5 Isolation in MIMO Configurations

In MIMO systems, isolation between antenna elements is a key performance metric. Poor isolation leads to coupling, which reduces the data rate and increases interference.

Isolation is measured by the mutual coupling parameter S_{12} (or S_{21}). For good MIMO performance, isolation should be less than −15 dB.

FSS can improve isolation by acting as a spatial filter or decoupling structure. For example, placing FSS between antenna elements reduces surface-wave coupling.

In mm-wave MIMO, isolation is harder to achieve because antennas are packed very closely to save space. Careful FSS design helps to maintain high isolation while keeping the array compact.

8.6.6 Trade-offs Between Complexity, Cost, and Performance

Finally, every design comes with trade-offs. A reconfigurable antenna with FSS may give high gain, wide bandwidth, and reconfigurability. But it also adds complexity in design, biasing, and fabrication.

At mm-wave, these trade-offs are even stronger. For example:

- adding MEMS switches improves performance but increases cost;
- using graphene enables wide reconfigurability but requires advanced fabrication;
- complex FSS shapes give better filtering but are harder to make at mm-wave.

Therefore, performance metrics must be considered together. The final design depends on the application. For a mobile phone, low cost and compactness are more important than a very high gain. For satellite communication, high efficiency and reconfigurability are more important, even if the cost is high.

8.6.7 Summary

In summary, performance metrics for mm-wave reconfigurable antennas with FSS cover many aspects: impedance matching, bandwidth, gain, efficiency, isolation, and trade-offs. Designers must optimize these metrics depending on the application. Understanding the balance between them is key to building practical and effective mm-wave systems.

8.7 Applications

mm-Wave reconfigurable antennas integrated with FSS are gaining significance in many modern technologies. They are used in wireless communication, radar, imaging, and energy systems. Their main advantage is that they can provide high gain, wide bandwidth, and smart reconfigurability in a compact form. This section explains some of the main applications: 5G and 6G networks, satellite and space links, radar and imaging systems, wireless power transfer, and integration with RIS.

8.7.1 5G and 6G mm-Wave Wireless Communication Systems

The deployment of 5G networks has driven a significant demand for mm-wave technology. Frequencies such as 28, 37, and 39 GHz are already used for 5G, with future 6G systems expected to operate at even higher frequencies, potentially exceeding 100 GHz.

The main reason for using mm-wave in wireless networks is the availability of a large bandwidth. This helps to support high data rates, low latency, and massive connectivity. However, mm-wave signals suffer from high path loss and poor penetration. To solve this, antennas with beam steering and reconfigurability are needed.

FSS-based antennas can act as filters and reflectors, which improve gain and reduce interference. For example, an FSS superstrate placed above a phased array can shape the beam to focus on a user. This is useful for mobile base stations and small cells.

The Shannon capacity formula shows the importance of bandwidth:

$$C = B \log_2(1 + \mathrm{SNR}), \tag{8.21}$$

where C is capacity, B is bandwidth, and SNR is signal-to-noise ratio. Increasing B through mm-wave bands directly increases capacity.

In 6G, the use of intelligent surfaces and reconfigurable antennas will allow networks to dynamically adapt beams, support holographic communications, and enable ultra-reliable low-latency links.

8.7.2 Satellite and Space Communications

Satellites require high-gain, low-weight antennas to communicate over large distances. Traditional parabolic reflectors are bulky and hard to steer quickly. At mm-wave, reconfigurable antennas with FSS provide a lighter and more flexible solution.

For example, an FSS reflector can act as a flat metasurface mirror, focusing energy without needing a large dish. Reconfigurable elements allow dynamic beam steering to track satellites or ground stations.

In low earth orbit (LEO) constellations, thousands of satellites are being launched to provide internet across the globe. These systems need compact, electronically steerable antennas that can reconfigure quickly. FSS-based solutions can meet this need while reducing power consumption and size.

Challenges include radiation hardness (resistance to cosmic radiation), thermal expansion in space, and ensuring that reconfigurable components like MEMS or graphene survive extreme conditions.

Equation for free-space path loss (FSPL) explains why a high gain is needed:

$$L_{fs} = \left(\frac{4\pi d f}{c}\right)^2, \tag{8.22}$$

where d is distance, f is frequency, and c is the speed of light. For large d (thousands of km) and high f (tens of GHz), L_{fs} becomes very high, demanding strong antennas.

8.7.3 mm-Wave Radar and Imaging Systems

Radar and imaging systems are another important field where mm-wave reconfigurable antennas with FSS are applied.

8.7.3.1 Security and Surveillance

At airports and other sensitive places, mm-wave imaging is used for body scanners. These systems rely on antennas that can scan beams and provide high resolution. Reconfigurable FSS layers can improve the resolution by controlling the bandwidth and polarization.

8.7.3.2 Automotive Radar

Modern vehicles use mm-wave radars at 24, 60, and 77 GHz for adaptive cruise control, collision avoidance, and parking assistance. Reconfigurable antennas allow better detection range and angular resolution. FSS layers can act as filters to reduce interference from nearby cars and improve isolation.

The radar range equation highlights the importance of the antenna gain:

$$R_{\max} = \left(\frac{P_t G_t G_r \lambda^2 \sigma}{(4\pi)^3 S_{\min}} \right)^{1/4}, \tag{8.23}$$

where P_t is transmitted power, G_t and G_r are transmitter and receiver gains, λ is wavelength, σ is radar cross section, and $S_{\min}$ is minimum detectable signal. Higher G_t and G_r (gain) allow longer detection range.

8.7.3.3 Biomedical Imaging

In medicine, mm-wave imaging is applied for cancer detection, skin imaging, and monitoring of vital signs. Reconfigurable antennas help to focus beams on targeted body parts and adjust frequency for better penetration depth. FSS superstrates can increase sensitivity and reduce noise.

8.7.4 Wireless Power Transfer and Energy Harvesting

Another growing application of mm-wave antennas with FSS is in wireless power transfer (WPT) and energy harvesting.

8.7.4.1 Wireless Power Transfer

WPT at mm-wave can provide energy to small sensors, implants, or IoT devices. The antenna must exhibit high efficiency to transfer energy over short to medium distances. FSS superstrates can focus the energy beam and improve efficiency.

The power received in WPT follows Friis transmission equation:

$$P_r = P_t G_t G_r \left(\frac{\lambda}{4\pi d} \right)^2, \tag{8.24}$$

where P_r is received power, P_t is transmitted power, and d is distance. At mm-wave, λ is small, so high G_t and G_r are needed.

8.7.4.2 Energy Harvesting

Energy harvesting at mm-wave is also possible using rectennas (rectifying antennas). These devices convert EM waves into DC power. FSS can improve harvesting efficiency by filtering desired frequencies and rejecting noise.

Such systems can be used in powering IoT sensors in smart cities or medical implants inside the human body.

8.7.5 Integration with RIS for Smart Environments

Reconfigurable intelligent surfaces are becoming a major part of future 6G systems. RIS panels are large surfaces made of many passive or semi-passive unit cells that can control reflection, refraction, or absorption of EM waves.

When combined with FSS, RIS can provide selective frequency control and polarization manipulation. This integration allows buildings, walls, and even furniture to act as smart communication nodes. For example:

- in an indoor environment, RIS with FSS can redirect beams to users in non-line-of-sight (NLOS) areas;
- in outdoor environments, RIS panels on building facades can enhance coverage and reduce shadowing; and
- in MIMO systems, RIS with FSS can help to improve isolation between antennas and reduce interference;

Challenges here include designing low-power biasing networks, scaling up to thousands of elements, and maintaining low cost.

8.7.6 Summary

Applications of mm-wave reconfigurable antennas with FSS cover many fields: mobile networks, satellite communications, radar, imaging, energy, and smart environments. Each application demands specific performance metrics such as high gain, efficiency, reconfigurability, and compact size. The combination of FSS and reconfigurable antennas provides flexible solutions to meet these demands, making them a key technology for future wireless and sensing systems.

8.8 Challenges and Future Trends

The development of mm-wave reconfigurable antennas with FSS faces many challenges. While the technology shows great promise for 5G, 6G, satellite, radar, and energy applications, there are still practical limitations in fabrication, thermal handling, integration, and material selection. At the same time, new trends such as tunable 2D materials and AI-driven optimization are opening exciting possibilities.

8.8.1 Fabrication and Tolerance Issues at mm-Wave Bands

Fabrication at mm-wave frequencies is one of the most critical challenges. At these high frequencies, the wavelength is very small (from 1 mm to 10 mm), so the FSS unit cells and antenna elements must be fabricated with micrometer-level precision. Even small deviations in dimensions can lead to significant frequency shifts or performance degradation.

For example, consider a square patch FSS unit cell with length L. A small error ΔL shifts the resonance frequency f_r approximately as:

$$\Delta f_r \approx -\frac{c}{2L^2\sqrt{\epsilon_{eff}}}\Delta L, \tag{8.25}$$

where c is the speed of light and ϵ_{eff} is the effective permittivity. This shows that even a few micrometers of error can move the resonance by hundreds of MHz at mm-wave, which is unacceptable for high-precision applications.

Other issues include alignment errors in multilayer structures, roughness of metallic surfaces, and nonuniformity in dielectric substrates. Advanced lithography, laser machining, and high-resolution 3D printing can be used to reduce these errors, but they increase the cost.

8.8.2 Thermal and Power Handling Constraints of Reconfigurable FSS

Reconfigurable FSS antennas often include active components such as PIN diodes, varactors, MEMS switches, or phase-change materials, which introduce additional power handling limitations.

At mm-wave, even small resistive losses in the metal or dielectric layers generate heat. The temperature rise can affect the biasing network and even damage active components. Thermal analysis is therefore critical. The temperature rise ΔT can be estimated as:

$$\Delta T = \frac{P_{\text{loss}}}{hA}, \tag{8.26}$$

where P_{loss} is the power dissipated, h is the heat transfer coefficient, and A is the surface area of the substrate.

Designers must ensure proper heat spreading, choose low-loss substrates, and sometimes include cooling solutions in compact modules. Power handling limits also restrict the maximum transmitted power in mm-wave systems, potentially causing issues in radar and satellite applications.

8.8.3 Integration with CMOS and Compact Packaging

For practical applications, FSS-based antennas must be integrated with CMOS chips, transceivers, and compact packages. At mm-wave frequencies, the integration becomes challenging because of the parasitic capacitance and inductance of packaging and interconnects.

For example, the input impedance Z_{in} of an integrated antenna can be affected by the parasitic reactance X_p of interconnects:

$$Z_{\text{in}} = R_{\text{antenna}} + j(X_{\text{antenna}} + X_p), \tag{8.27}$$

If X_p is not properly managed, matching is lost and the reflection coefficient S_{11} increases.

Compact packaging also introduces thermal challenges and limits the space available for biasing networks of reconfigurable elements. Multilayer PCB designs and through-silicon vias (TSVs) are often used, but these increase fabrication complexity.

8.8.4 Emergence of Tunable 2D Materials for FSS-based Reconfigurable Antennas

New materials are emerging that can make FSS-based antennas more efficient and tunable. Materials such as graphene, molybdenum disulfide (MoS_2), and vanadium dioxide (VO_2) show tunable electrical properties under voltage, temperature, or optical control.

8.8.4.1 Graphene

Graphene has high electron mobility and can support plasmonic waves at mm-wave and THz frequencies. Its surface conductivity σ_s can be tuned with a bias voltage V_g:

$$\sigma_s = \frac{e^2 E_F}{\pi \hbar^2 (\gamma - j\omega)}, \tag{8.28}$$

where E_F is the Fermi level, $\hbar$ is the reduced Planck constant, γ is scattering rate, and ω is angular frequency. This tunability allows dynamic control of resonance, bandwidth, and polarization of FSS structures.

8.8.4.2 VO_2 and MoS_2

VO_2 can switch from an insulator to a conductor at around 68 °C, which allows temperature-controlled reconfigurability. MoS_2 is useful for voltage-controlled FSS designs. Using these materials, antennas can adapt to environmental changes, enabling smart wireless systems.

8.8.5 AI/ML-driven Optimization for Adaptive FSS-based Antenna Systems

Another future trend is the use of artificial intelligence (AI) and machine learning (ML) for antenna and FSS design. Traditional design requires multiple simulations and parametric sweeps, which is time-consuming. AI/ML algorithms can learn patterns from previous designs and predict optimal parameters quickly.

For example, a neural network can be trained to predict S_{11} and S_{21} of an FSS unit cell given geometry, material, and frequency. Once trained, it can optimize multiband, multi-polarization, or wideband designs in seconds.

AI can also help in adaptive systems where FSS antennas adjust in real-time to changing environments. Reinforcement learning can be used to adjust phase, polarization, or frequency to maximize signal-to-noise ratio (SNR) or coverage.

8.8.6 Future Perspective: Convergence of FSS, RIS, and mm-Wave Antennas for 6G and THz Systems

Looking ahead, the convergence of FSS, RIS, and mm-wave/THz antennas is a major trend for 6G and beyond.

RIS can manipulate waves in real-time over large surfaces. By integrating tunable FSS with RIS panels, it is possible to create smart environments where signals are controlled dynamically for optimal coverage, high data rates, and energy efficiency.

At THz frequencies (0.1–10 THz), antenna size becomes extremely small, making FSS and metasurfaces even more important for beam shaping, multiband operation, and polarization control. AI-driven optimization will play a central role in managing complex systems with thousands of tunable elements.

8.8.7 Summary

In summary, the challenges of mm-wave FSS-based reconfigurable antennas include fabrication precision, thermal handling, compact integration, and active component limitations. Future trends point toward advanced materials, AI-based design and control, and integration with RIS for smart 6G/THz networks. Overcoming these challenges while leveraging new technologies will enable highly efficient, adaptive, and compact mm-wave systems for next-generation wireless communication, sensing, and energy applications.

8.8.8 Summary of the Role of FSS in Enabling Reconfigurability

FSS have shown remarkable capabilities in controlling EM waves at mm-wave frequencies. By using carefully designed unit cells arranged in periodic structures, FSS can filter, reflect, or transmit waves selectively. When combined with reconfigurable elements such as PIN diodes, varactors, MEMS, and tunable 2D materials like graphene and VO_2, FSS allows antennas to adapt dynamically to different operating conditions.

Reconfigurability is achieved in several aspects:

- **Frequency Reconfigurability:** FSS allows antennas to shift their operating frequency to cover multiple bands without changing physical size. This is important for 5G/6G networks and multi-standard devices.

- **Polarization Reconfigurability:** By modifying the geometry or activating specific elements, the polarization of the antenna can be changed from linear to circular or rotated by a desired angle. This improves link reliability and reduces interference.
- **Beam and Gain Control:** FSS superstrates can focus or steer beams, increasing antenna gain and directing energy toward desired directions. This is critical for high-frequency mm-wave and THz communications where path loss is high.
- **Multi-band and Wideband Operation:** Complex FSS geometries, such as fractal or multi-resonant unit cells, allow antennas to operate across several bands simultaneously, which is essential for versatile wireless and sensing systems.

The combination of FSS and reconfigurable antennas reduces the need for mechanical movement, improves system efficiency, and enables compact designs suitable for mobile, satellite, and IoT applications.

8.8.9 Potential for Next-generation Wireless and Sensing Systems

The capabilities of mm-wave reconfigurable antennas with FSS make them particularly promising for next-generation technologies:

8.8.9.1 5G and 6G Networks

High-speed wireless communication requires wide bandwidth, high data rates, and low latency. FSS-based reconfigurable antennas allow dynamic frequency tuning, beam steering, and interference reduction. They can be deployed in small cells, massive MIMO arrays, and smart base stations to enhance coverage and spectral efficiency.

8.8.9.2 Satellite and Space Communication

For satellite links, these antennas provide high-gain, compact solutions capable of beam steering without large mechanical reflectors. Their adaptability allows communication with multiple ground stations and dynamic satellite constellations in LEO or geostationary orbit (GEO).

8.8.9.3 Radar and Imaging Applications

In radar and imaging systems, FSS-based reconfigurable antennas improve angular resolution, scanning flexibility, and frequency agility. They are used in automotive radar, security scanners, and biomedical imaging systems, providing enhanced detection and image quality.

8.8.9.4 Energy Harvesting and Wireless Power Transfer

The ability to control beam direction and frequency allows efficient wireless power transfer and energy harvesting. FSS can filter desired frequencies and direct energy toward devices or sensors, making them suitable for IoT networks, implantable medical devices, and smart city applications.

8.8.9.5 Integration with Smart Environments

Reconfigurable FSS can be integrated with RIS to create smart environments. For example, walls, ceilings, and furniture can dynamically control wireless signals, improving coverage, reducing interference, and increasing energy efficiency. This approach is essential for future 6G networks and beyond.

8.8.10 Open Research Directions

Despite the progress, several research challenges remain, offering opportunities for future exploration.

8.8.10.1 Advanced Materials

The use of tunable 2D materials, such as graphene, MoS_2, and VO_2, is still in early stages. Research is needed to improve fabrication, stability, and large-area implementation. Understanding the long-term reliability and thermal behavior of these materials in practical systems is also important.

8.8.10.2 Ultra-compact Integration

Integrating FSS-based antennas with CMOS chips and packaging for compact devices remains challenging. Future research should focus on reducing parasitics, improving thermal management, and creating cost-effective integration methods for mass production.

8.8.10.3 AI and Machine Learning for Adaptive Control

The design and real-time control of FSS antennas can benefit greatly from AI and ML. Research into reinforcement learning, neural network-based optimization, and adaptive beamforming can enable antennas that automatically adjust to changing environments and user requirements.

8.8.10.4 High-frequency (THz) Applications

As wireless systems move toward THz frequencies, FSS design becomes more complex due to extremely small wavelengths and fabrication tolerances. Future work should focus on accurate modeling, material innovation, and high-precision manufacturing to make THz reconfigurable antennas feasible for practical applications.

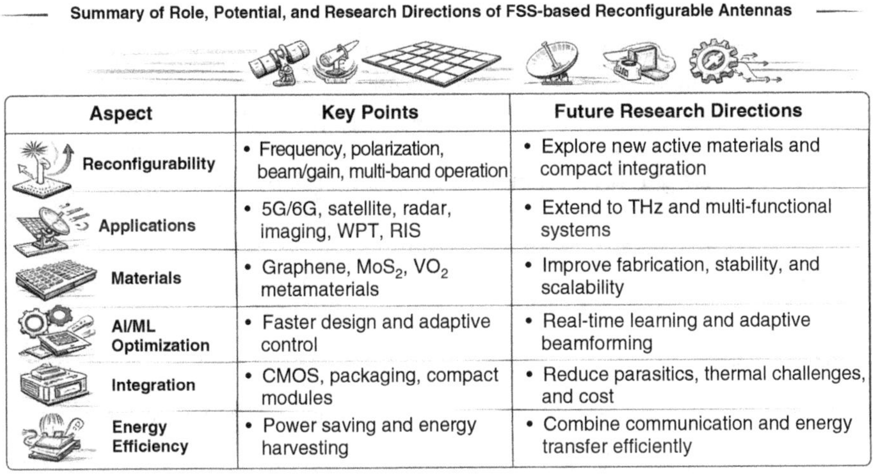
Summary of Role, Potential, and Research Directions of FSS-based Reconfigurable Antennas

Aspect	Key Points	Future Research Directions
Reconfigurability	• Frequency, polarization, beam/gain, multi-band operation	• Explore new active materials and compact integration
Applications	• 5G/6G, satellite, radar, imaging, WPT, RIS	• Extend to THz and multi-functional systems
Materials	• Graphene, MoS_2, VO_2 metamaterials	• Improve fabrication, stability, and scalability
AI/ML Optimization	• Faster design and adaptive control	• Real-time learning and adaptive beamforming
Integration	• CMOS, packaging, compact modules	• Reduce parasitics, thermal challenges, and cost
Energy Efficiency	• Power saving and energy harvesting	• Combine communication and energy transfer efficiently

Figure 8.11 Summary of key aspects.

8.8.10.5 Energy Efficiency and Sustainable Design

Next-generation systems must be energy efficient. Research is needed to design FSS-based antennas that maximize performance while minimizing power consumption. Combining energy harvesting and wireless communication in a single system is also a promising direction.

8.8.10.6 Multi-functional Antennas

FSS-based antennas that can perform multiple functions, such as communication, sensing, radar, and energy harvesting, are a key research area. Optimizing designs to achieve multifunctionality without sacrificing performance or efficiency remains a challenge.

8.8.11 Summary Table of Key Points

A summary of key aspects is shown in Figure 8.11.

8.9 Conclusion

In conclusion, FSS-enabled reconfigurable antennas provide a versatile, high-performance, and compact solution for mm-wave and THz systems. They enable frequency, polarization, and beam reconfigurability, which are essential for modern communication, sensing, and energy applications.

The integration of tunable materials, AI-driven optimization, and RIS-based smart environments is likely to drive the next generation of wireless systems. Open research directions include material innovation, ultra-compact integration, THz applications, energy-efficient designs, and multifunctional antennas.

As the demand for high-speed, reliable, and adaptive wireless systems grows, FSS-based reconfigurable antennas will play a central role in enabling the technology of the future. With continued research and development, these antennas have the potential to transform wireless communication, radar, imaging, and energy transfer systems, making them smarter, faster, and more efficient.

References

Saidatul. Norlyana. Azemi, Kamran. Ghorbani, and Wayne. S.T. Rowe. A reconfigurable FSS using a spring resonator element. *IEEE Antennas and Wireless Propagation Letters*, 12:781–784, 2013. https://doi.org/10.1109/LAWP.2013.2270950.

Hao Chen, Wei-Bing Lu, Zhen-Guo Liu, and Ming-Yang Geng. Microwave programmable graphene metasurface. *ACS Photonics*, 7(6):1425–1435, 2020.

Alireza Foroozesh and Lotfollah Shafai. Investigation into the effects of the patch-type FSS superstrate on the high-gain cavity resonance antenna design. *IEEE Transactions on Antennas and Propagation*, 58(2):258–270, 2010. https://doi.org/10.1109/TAP.2009.2037702.

María Garcia-Vigueras, José Luis Gomez-Tornero, George Goussetis, Juan Sebastian Gomez-Diaz, and Alejandro Alvarez-Melcon. A modified pole-zero technique for the synthesis of waveguide leaky-wave antennas loaded with dipole-based FSS. *IEEE Transactions on Antennas and Propagation*, 58(6):1971–1979, 2010. https://doi.org/10.1109/TAP.2010.2046856.

Lei Gong, Xinru Tian, Ge Zhang, Xuan Liu, Shigeng Song, Jian Song, Haoyang Liu, Liguo Wang, and Zhiqiang Yang. AFSS wide-frequency reconfigurable design and electromagnetic characterization research. *Electronics*, 14(18):3628, 2025.

Suguna Gunasekaran, Manikandan Chinnusami, Rajesh Anbazhagan, Karunyaa Sureshkumar, and Shreela Sridhar. Multi-band unmanned aerial vehicle antenna for integrated 5G and GNSS connectivity. In *Telecom*, vol. 6, page 38. MDPI, 2025.

Mehdi Hosseini and Mohammad Hakkak. Characteristics estimation for Jerusalem cross-based artificial magnetic conductors. *IEEE Antennas and Wireless Propagation Letters*, 7:58–61, 2008. https://doi.org/10.1109/LAWP.2008.917605.

Mirshahram Hosseinipanah and Qun Wu. Novel frequency selective surfaces designing by modification of Jerusalem cross slot. In *2008 International Conference on Microwave and Millimeter Wave Technology*, vol. 1, pages 44–47, 2008. https://doi.org/10.1109/ICMMT.2008.4540296.

Hao Huang, Changzhou Hua, and Zhongxiang Shen. Absorptive frequency-selective transmission structures based on hybrid FSS and absorber. *IEEE Transactions on*

Antennas and Propagation, 70(7):5606–5613, 2022. https://doi.org/10.1109/TAP.2022.3161472.

Jieun Hwang and Sungcheol Hong. Passive frequency tunability in moiré-inspired frequency selective surfaces based on full-wave simulation. *Micromachines*, 16(6):702, 2025.

Lu-Yang Ji, Zhi-Ya Zhang, and Neng-Wu Liu. A two-dimensional beam-steering partially reflective surface (PRS) antenna using a reconfigurable FSS structure. *IEEE Antennas and Wireless Propagation Letters*, 18(6):1076–1080, 2019. https://doi.org/10.1109/LAWP.2019.2907641.

Anand Kumar, Saikiran Kongari, Yugesh Chandrakapure, and Debdeep Sarkar. Multi-functional metasurface as a transmissive/reflective FSS and an on-air frequency mixer. *Scientific Reports*, 14(1):13874, 2024.

Bin Liang, Benito Sanz-Izquierdo, Edward A. Parker, and John C. Batchelor. Cylindrical slot FSS configuration for beam-switching applications. *IEEE Transactions on Antennas and Propagation*, 63(1):166–173, 2015. https://doi.org/10.1109/TAP.2014.2367534.

Hanxiang Lin and Xiaoxing Fang. An improved PIN diode model design for a tunable frequency selective absorber. *Applied Sciences*, 15(3):1440, 2025.

Linshuang Long, Sydney Taylor, and Liping Wang. Enhanced infrared emission by thermally switching the excitation of magnetic polariton with scalable microstructured VO_2 metasurfaces. *ACS Photonics*, 7(8):2219–2227, 2020.

Zhongliang Lv, Zelun Li, Yu Han, Yujiang Cao, and Lin Yang. A wideband and ultra-thin metamaterial absorber based on resistive FSS. *Symmetry*, 14(6):1148, 2022.

S. Maci, M. Caiazzo, A. Cucini, and M. Casaletti. A pole-zero matching method for EBG surfaces composed of a dipole FSS printed on a grounded dielectric slab. *IEEE Transactions on Antennas and Propagation*, 53(1):70–81, 2005. https://doi.org/10.1109/TAP.2004.840520.

C Mias and C Tsakonas. Waveguide demonstration of varactor-diode-tunable band-pass frequency-selective surface. *Microwave and Optical Technology Letters*, 45(1):62–66, 2005.

Yaxi Pan and Jian Dong. Design and optimization of an ultrathin and broadband polarization-insensitive fractal FSS using the improved bacteria foraging optimization algorithm and curve fitting. *Nanomaterials*, 13(1):191, 2023.

Akash Paramanik, Krishnamurthy Chitti Venkata, and Venkatachalam Subramanian. A dual layer wideband angular stable frequency selective surface for linear to circular and circular to linear polarization conversion for 5G applications. *Scientific Reports*, 15(1):12059, 2025.

J. Romeu and Y. Rahmat-Samii. A fractal based FSS with dual band characteristics. In *IEEE Antennas and Propagation Society International Symposium. 1999 Digest. Held in conjunction with: USNC/URSI National Radio Science Meeting (Cat. No. 99CH37010)*, vol. 3, pages 1734–1737, 1999. https://doi.org/10.1109/APS.1999.788289.

B. Schoenlinner, A. Abbaspour-Tamijani, L.C. Kempel, and G.M. Rebeiz. Switchable low-loss RF mems ka-band frequency-selective surface. *IEEE Transactions on Microwave Theory and Techniques*, 52(11):2474–2481, 2004. https://doi.org/10.1109/TMTT.2004.837148.

Yogesh Solunke and Ashwin Kothari. A low-RCS dual-bandstop golden ratio-based fractal-FSS for defense applications. *International Journal of Communication Systems*, 38(4):e5997, 2025.

Chao Sun, Guangyi Heng, Yuhang Zou, Dongmin Zhang, Chen Chen, and Jiahui Fu. A miniaturized FSS using the parallel LC resonant with angular stability. *Sensors (Basel, Switzerland)*, 25(16):4931, 2025.

Riku Tanaka, Hiroya Hayakawa, Yudai Handa, Jaesang Cha, and Chang-Jun Ahn. Advanced technology in ATSC3. 0: Boosting data signal rates using polarization properties and FSS. *Applied Sciences*, 15(11):6306, 2025.

Hassan Umair, Tarik Bin Abdul Latef, Yoshihide Yamada, Wan Nor Liza Binti Wan Mahadi, Mohamadariff Othman, Kamilia Kamardin, Mousa I Hussein, and Ali Imran Najam. Tilted beam Fabry–Perot antenna with enhanced gain and broadband low backscattering. *Electronics*, 10(3):267, 2021.

Batuhan Uslu, Sena Esen Bayer Keskin, and Nurhan Türker Tokan. Dual-mode reconfigurable frequency-selective surface for switching between narrowband and wideband applications. *Micromachines*, 16(9):1030, 2025.

Jingbo Wu, Ze Shen, Shijun Ge, Benwen Chen, Zhixiong Shen, Taofeng Wang, Caihong Zhang, Wei Hu, Kebin Fan, Willie Padillaet al. Liquid crystal programmable metasurface for terahertz beam steering. *Applied Physics Letters*, 116(13): 131104, 2020.

Mingbao Yan, Jiafu Wang, Yongqiang Pang, Cuilian Xu, Hongya Chen, Lin Zheng, Jieqiu Zhang, and Shaobo Qu. An FSS-backed dual-band reflective polarization conversion metasurface. *IEEE Access*, 7:104435–104442, 2019. https://doi.org/10.1109/ACCESS.2019.2931551.

Rui Zhao, Boyi Gong, Fajun Xiao, Chong He, and Weiren Zhu. Circuit model analysis of switchable perfect absorption/reflection in an active frequency selective surface. *IEEE Access*, 7:55518–55523, 2019. https://doi.org/10.1109/ACCESS.2019.2913406.

Hang Zhou, Liwei Yang, Shaobo Qu, Kai Wang, Jiafu Wang, Hua Ma, and Zhuo Xu. Experimental demonstration of an absorptive/transmissive FSS with magnetic material. *IEEE Antennas and Wireless Propagation Letters*, 13:114–117, 2014. https://doi.org/10.1109/LAWP.2013.2296992.

Yassine Zouaoui, Larbi Talbi, Khelifa Hettak, and Naresh K. Darimireddy. Synthesis of planar and conformal single-layered double-sided parallel-cross dipole FSS based on closed-form expression. *IEEE Access*, 9:104051–104058, 2021. https://doi.org/10.1109/ACCESS.2021.3096419.

9

Orbital Angular Momentum Antennas

9.1 Introduction

The growing demand for next-generation wireless communication systems has pushed researchers to explore unconventional dimensions of electromagnetic waves. Orbital angular momentum (OAM), an intrinsic characteristic of electromagnetic radiation with a helical phase, has emerged as a powerful tool to enhance channel capacity, enable mode multiplexing, and introduce new paradigms for secure and high-capacity links. This chapter provides a detailed discussion of the theoretical foundations of OAM, practical antenna designs for OAM generation at millimeter-wave (mm-wave) frequencies, key challenges such as mode purity and divergence, and promising applications, including high-capacity backhaul, secure links, and radar. We also discuss future perspectives, including OAM-enabled sixth-generation (6G) systems.

Every new generation of communication technology brings us closer to a world where billions of devices are seamlessly connected and exchange information without interruption. Because of this trend, researchers have been actively exploring innovative ideas beyond conventional solutions. One of several emerging areas for addressing this trend is using OAM waves to form secure antennas.

Wireless communication has relied heavily on conventional approaches such as frequency division, time division, polarization diversity, and pattern diversity. These techniques have been highly successful, but they are reaching a point where adding more users or increasing data rates becomes harder without causing interference. OAM provides a fresh degree of freedom because it allows the electromagnetic wave itself to carry extra "twist" in its phase. More specifically, the wave not only travels in the forward direction but also rotates around its axis. This special behavior of OAM makes it more promising as different twisting can travel together with no interference. Each twist (mode) can carry information; thus, they can enhance the data capacity.

OAM antennas are generally different from traditional antennas. Traditional antennas are designed to control frequency, amplitude, and polarization. On the other hand, OAM antennas are designed in a careful manner to generate a spiral phase front. This spiral nature is what makes them stand out. Some antennas achieve this by mechanical design, such as spiral phase plates (SPPs) or specially shaped reflectors, while others rely on advanced electronic structures like metasurfaces and phased arrays. Regardless of the method, the main aim is the same: to produce a wavefront that carries OAM.

The concept of OAM is not new in physics. It has been studied for many years in optics, where light beams with twisted wavefronts are used in laser communication, imaging, and even in trapping microscopic particles. What is new is its use at radio and mm-wave frequencies, such as fifth-generation (5G), 6G, and beyond. mm-wave bands are attractive because they offer large amounts of unused spectrum. However, they also face challenges like high path loss and limited coverage. By combining mm-wave communication with OAM, researchers hope to not only expand capacity but also improve the efficiency of short-range, high-speed links.

One way to imagine OAM is to picture water flowing in a river. In typical scenarios, the water flows straight along the path. But if you stir the water with a stick, whirlpools appear, and the flow gains a twisting motion. This is similar to what happens in electromagnetic waves when OAM is introduced. Instead of just flowing forward, the energy swirls around its axis. Each swirl, known as a mode, can be thought of as an independent channel. So, by assigning different modes to different data streams, multiple users can communicate at the same time without interfering with each other.

OAM antennas represent a new concept in wireless communication; however, several challenges must be addressed for them to be practically useful. First, efficiency is important. If the antenna is not designed well, the twisted wave can lose quality and reduce performance. Another problem is alignment. Since OAM beams spread and have a hollow center, even a small misalignment between the transmitter and receiver can cause issues. Researchers are actively exploring reconfigurable antennas and adaptive methods to fix these problems.

In modern communication systems, OAM antennas are not only famous for enhancing wireless links by increasing the channel capacity, but also for improving wireless communication links, imaging, radar, and satellite data. In highly secure communication systems, OAM antennas can add an extra layer of security. In such systems, the receiver should know the exact mode to decode the received waves. In the case of radar and imaging, they improve image and target resolution. In satellite communication, they reduce the spectrum overloading. Even having many advantages, OAM is yet to be considered a complete solution. It can be used with other technologies to increase its scope and utilities. Such technologies include, but are not limited to MIMO, beamforming, and reconfigurable intelligent surfaces (RIS).

To summarize, OAM antennas introduce new opportunities for highly secure and high-speed wireless communication for variety of applications. Along with attractive applications, they need new designs and smart engineering in the mm-wave frequency range. Lots of research is being conducted on these antennas, as they are a hot topic. In fact, researchers consider it an important part of next-generation wireless systems.

9.2 Fundamentals of OAM

9.2.1 Angular Momentum in Electromagnetics

EM waves are not just simple carriers of energy; they can also carry different types of angular momentum. There are two main forms: spin angular momentum (SAM) and OAM.

- **SAM** is linked to the wave's polarization. If the wave is circularly polarized, the electric field rotates either clockwise or counterclockwise as the wave travels. These two states correspond to spin values of $+\hbar$ or $-\hbar$ per photon, where $\hbar$ is the reduced Planck constant.
- **OAM**, on the other hand, comes from the way the wavefront twists as it propagates. Unlike polarization, which only has two possible states (left or right circular), OAM can take on many different states, labeled by an integer called the *mode index* ℓ. This means OAM offers a theoretically unlimited number of distinct channels.

The mathematical form of an OAM beam is usually written as:

$$E(r, \phi, z) = A(r, z)e^{j(\ell\phi + kz - \omega t)}, \tag{9.1}$$

where

- $A(r, z)$ describes how strong the field is at a given radius r and position z.
- ℓ is the OAM mode index, which tells us how many times the wavefront twists around its axis in one wavelength.
- ϕ is the azimuthal angle around the beam axis.
- $kz - \omega t$ represents the usual forward-traveling wave.

The key part is the factor $e^{j\ell\phi}$. This term gives the wavefront a *helical or corkscrew-like shape*. When $\ell = 1$, the wave twists once per cycle; when $\ell = 2$, it twists twice, and so on. This twist creates a *doughnut-shaped intensity profile* with a hollow center, because the phase singularity at the axis forces the field there to vanish.

In simple words, SAM is about the "spin" of the wave's polarization, while OAM is about the "twist" of its wavefront. Together, they make electromagnetic waves much richer than just simple plane waves, opening up new ways to carry information and interact with matter.

9.2.2 Orthogonality of OAM Modes

One of the most important properties of OAM modes is their orthogonality. This means that different OAM modes do not interfere with each other, even when they travel along the same path. Mathematically, this can be expressed as:

$$\int_0^{2\pi} e^{j\ell_1\phi}e^{-j\ell_2\phi}d\phi = 0, \quad \ell_1 \neq \ell_2. \tag{9.2}$$

Here, ℓ_1 and ℓ_2 are two different OAM mode indices. If the indices are not equal, the integral evaluates to zero, showing that the modes are orthogonal. In simple words, each mode has a unique "twist" pattern that does not overlap with others.

This property is very useful for communication systems. Because OAM modes do not interfere, we can use each mode as a separate data channel. This allows transmission of multiple streams of information over the same band and direction, effectively increasing the data capacity of a communication link. In practical terms, it is like having several separate highways for data, all occupying the same physical space without collisions.

Orthogonality also helps in signal recovery at the receiver. By knowing which OAM mode corresponds to which data stream, the receiver can accurately extract the information without confusion from other modes. This makes OAM a powerful tool for high-capacity and efficient wireless communication.

9.3 OAM Generation

OAM beams need special antennas to generate them. These antennas shape the electromagnetic wave so that it twists around its axis. There are several methods to create OAM beams. Here, we explain the main techniques in a simple language. Key methods for generating OAM modes are shown in Figure 9.1.

9.3.1 Spiral Phase Plates

SPPs are simple devices that twist a wave. They have a spiral shape, and when a wave passes through, the thickness changes with the angle. This change in thickness makes the wavefront rotate, creating OAM. SPPs are easy to make using 3D printing or machining. The main disadvantage is that they cannot easily change the OAM mode once built.

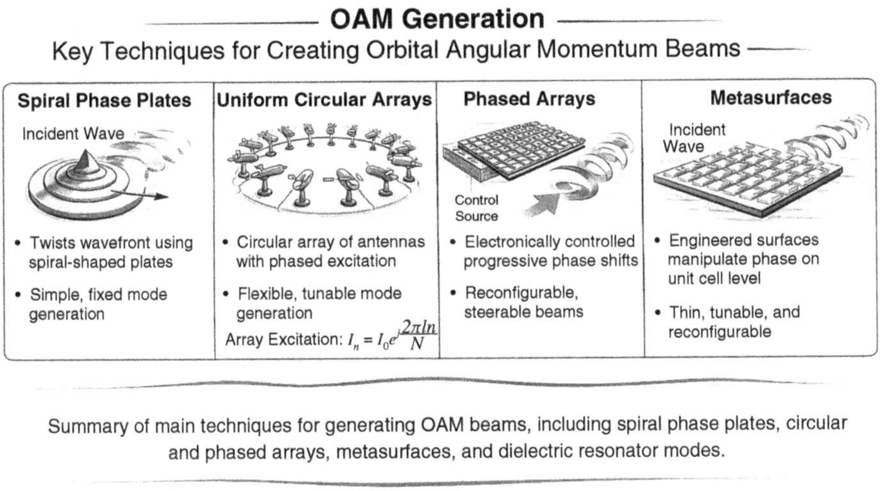

Figure 9.1 Key methods for generating OAM modes.

9.3.2 Uniform Circular Arrays

A uniform circular array (UCA) is a group of antennas arranged in a circle. Each antenna is fed with the same signal but with a small difference in phase. By carefully choosing this phase difference, the combined wave from all antennas twists around the center. This method can generate different OAM modes by changing the phase. UCAs are flexible and allow beam steering.

The excitation for each element in a UCA can be written as:

$$I_n = I_0 e^{j\frac{2\pi \ell n}{N}}, \quad n = 0, 1, 2, \ldots, N-1, \tag{9.3}$$

where I_0 is the amplitude, N is the number of antennas, and ℓ is the OAM mode.

9.3.3 Phased Arrays with Progressive Phase Shift

Phased arrays are antennas where each element can control its phase electronically. To generate an OAM beam, each element is given a phase shift that changes progressively around the array. This makes the total wave twist like a corkscrew. Phased arrays offer significant advantages, as they allow dynamic changes to the OAM mode and real-time beam steering.

9.3.4 Metasurfaces

Metasurfaces are thin layers made of tiny engineered structures called unit cells. Each unit cell can change the phase of a wave that passes through it. By arranging the unit cells properly, a helical phase front is created. Metasurfaces can be made to generate different OAM modes. Reconfigurable metasurfaces allow switching between modes by electronically controlling the unit cells. This makes them useful for adaptive systems.

9.3.5 Dielectric Resonator Modes

Dielectric resonators are small blocks of material that naturally support electromagnetic modes. Certain higher-order modes in these resonators have a twisted phase structure. By exciting these modes, the resonator can emit an OAM beam. Dielectric resonators are compact, efficient, and suitable for mm-wave applications.

9.3.6 Array Factor for OAM Modes

The array factor is a way to calculate how multiple antennas combine to form a desired wave. For a circular array generating OAM, the array factor can be written as:

$$\mathrm{AF}(\phi) = \sum_{n=0}^{N-1} I_n e^{jkr_n \cos(\phi-\phi_n)}, \tag{9.4}$$

where r_n and ϕ_n are the position of the nth element. By using the excitation $I_n = I_0 e^{j2\pi\ell n/N}$, the array factor forms a twisted wavefront, producing the OAM beam.

In summary, OAM beams can be generated using simple mechanical devices like SPPs, circular arrays with phase shifts, advanced phased arrays, engineered metasurfaces, and dielectric resonators. Each method has advantages and challenges, and the choice depends on the application, frequency, and flexibility required.

9.4 Challenges in OAM Antennas

While OAM antennas offer exciting opportunities, they also face several challenges. These challenges become more significant, especially at mm-wave frequencies, where the wavelength is small and the waves can be easily disturbed. Key challenges in OAM antennas are summarized in Figure 9.2.

9.4.1 Beam Divergence

One main challenge of OAM beams is that they spread out more quickly than normal waves. This spreading is called beam divergence. The higher the OAM

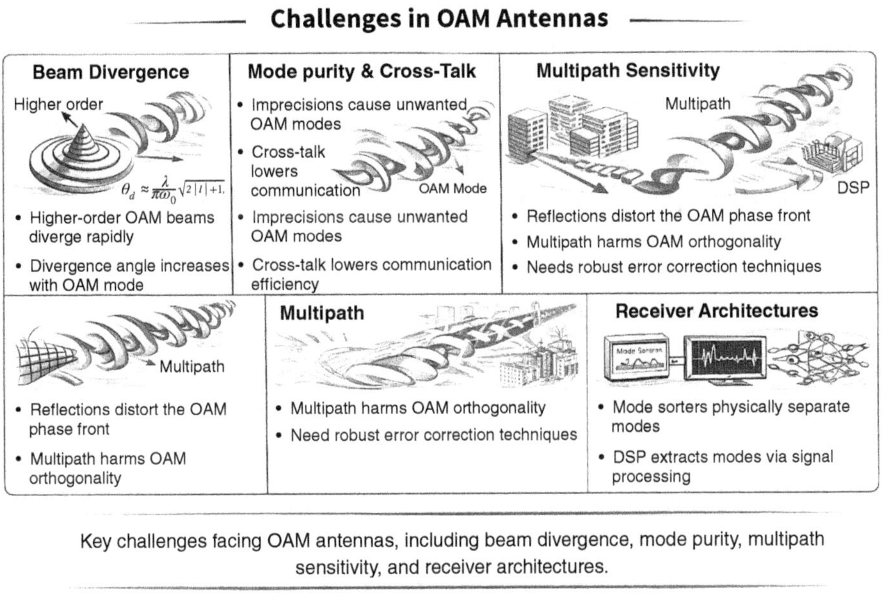

Figure 9.2 Key challenges in OAM antennas.

mode number ℓ, the faster the beam spreads. For a simple Gaussian beam carrying OAM, the divergence angle θ_d can be approximated as:

$$\theta_d \approx \frac{\lambda}{\pi w_0}\sqrt{2|\ell| + 1}, \tag{9.5}$$

where λ is the wavelength and w_0 is the beam waist. This equation shows that increasing ℓ makes the beam wider, which reduces the effective distance the signal can travel without losing strength. Designers need to balance the number of OAM modes with the link distance.

9.4.2 Mode Purity and Cross-talk

OAM antennas generate multiple modes, but sometimes unwanted modes appear due to imperfections in design or fabrication. This reduces mode purity. When modes are not pure, they can interfere with each other, causing cross-talk. Cross-talk decreases the efficiency of OAM communication because the data from one mode can mix with another. Careful antenna design and precise feeding networks are necessary to keep the modes clean and reduce interference.

9.4.3 Multipath Sensitivity

In real environments, electromagnetic waves often bounce off walls, buildings, and other objects. This is called multipath propagation. OAM beams are sensitive to these reflections because the twisting phase front can get distorted. Multipath can reduce the orthogonality of OAM modes, making it harder to separate the data channels. Researchers are studying ways to make OAM communication more robust in realistic channels, using error correction and adaptive beam shaping.

9.4.4 Receiver Architectures

To detect OAM beams, the receiver must correctly identify the mode. There are several types of receivers:

- **Mode Sorters:** Optical or RF devices that separate each OAM mode physically.
- **Digital Signal Processing (DSP):** Using algorithms to extract each mode from the received signal.
- **Machine Learning (ML)-based Demodulation:** Advanced AI methods can learn to detect OAM modes even in complex environments with noise and multipath.

Each receiver type has its advantages and challenges. Mode sorters are accurate but may be bulky. DSP is flexible but requires high computational power. ML methods are promising but need training and careful tuning.

In summary, while OAM antennas have great potential, they face challenges such as beam divergence, mode purity, sensitivity to multipath, and the need for sophisticated receivers. Overcoming these challenges is necessary to make OAM communication practical in real-world applications, especially for high-speed, high-capacity mm-wave systems.

9.5 Applications of OAM Antennas

OAM antennas are not only an area of active research but also have many practical applications. Their unique ability to carry multiple data streams on the same frequency makes them useful in different fields. Here, we present a discussion of the key applications in a simplified manner for clarity.

9.5.1 Point-to-point High-capacity Links

One of the most important uses of OAM antennas is in point-to-point communications. These links connect two locations directly and need high data rates.

OAM antennas can send multiple twisted modes at the same time, increasing the total capacity without using extra frequency bands. This makes them very useful for wireless backhaul in 5G and future 6G networks.

9.5.2 Secure Communications

OAM can also improve security in wireless systems. Because each OAM mode exhibits a unique phase structure, the receiver must know the exact mode to decode the signal. If an attacker does not know the mode, it becomes very hard to intercept the information. This property adds a layer of physical-layer security without relying only on encryption.

9.5.3 Chip-to-chip Wireless Links

In electronics, high-speed communication between chips is often required. OAM antennas at mm-wave frequencies can enable wireless links between chips on a board or in data centers. This avoids the need for long wires, reduces latency, and supports very high data rates, which is useful for cloud computing and high-performance computing systems.

9.5.4 Radar and Imaging Applications

OAM antennas can also improve radar and imaging systems. The twisting wavefront provides additional information about the target. This helps in separating multiple targets and improves resolution. For example, in radar, OAM modes can help distinguish closely spaced objects. In imaging, they can improve the clarity and detail of the images.

In summary, OAM antennas have many promising applications. They can provide high-capacity wireless links, enhance security, enable fast chip-to-chip communication, and improve radar and imaging systems. With ongoing research, these applications are likely to become more practical in the near future.

9.6 Future Outlook

Future directions for OAM antennas include:

- Integration with RIS to dynamically correct OAM distortions.
- Hybrid OAM-MIMO systems to combine spatial and modal multiplexing.
- Machine learning algorithms for adaptive beam detection and demultiplexing.
- Extending OAM to THz frequencies for ultra-broadband communication.

9.7 Design Example

Antennas with OAM beams can greatly increase data capacity and improve data security (Hui et al., 2015; Yan et al., 2014). This is possible because the modes created by the shared radiator are orthogonal. In other words, each mode of the radiator carries information with a unique OAM beam. With this feature, several information packets can be simultaneously delivered to a receiver using only one frequency band. However, to transmit such data packets simultaneously, an efficient antenna capable of generating many OAM modes is required. In Naseri et al. (2023), dual-polarized Huygens' metasurface realizes mixed modes in a transmitter. Two concentric TAs produce the OAM phase fronts by applying the correct phase shift to each unit cell according in its position. The dual-polarized Huygens' metasurface generates OAM −1 and −2 modes at 28 GHz. The TA has an overall size of 110×110 mm^2. A good gain of 16 dBi was achieved.

9.7.1 Literature Review

Based on existing literature, there are several antennas available that can efficiently generate OAM modes. These antenna radiators are grouped as (1) OAM antenna with one port, (2) OAM antennas where each port generates a distinct OAM mode, (3) an antenna with a wideband where each frequency point generates a different mode, and (4) OAM modes using metamaterials.

In Liu et al. (2018b), the authors have generated +1 and −1 OAM modes using four-element circular array configuration. Using similar type of antenna resonators, the authors in Li et al. (2020a) have generated +1 and −1 modes by designing different feeding network. In Liu et al. (2017), the authors change the direction of the input signal, which changes the individual signals' phases going towards radiator. This helps in creating reconfigurable modes. Another design in Kang et al. (2020) uses a special feed system that allows control over polarization and modes. In Li et al. (2020b), a metallic resonator is used to generate second-order OAM modes. A broadband antenna array is introduced in Huang and Zhang (2021), which enables the array to operate in dual-mode configuration. In Wu et al. (2019), −1, 0, and +1 modes are generated with the help of a metasurface antenna that can change modes electronically. The work in Kang et al. (2019) presents a simple feeding network that produces several modes. A 1-bit tunable array is used in Wang et al. (2020) to switch between OAM modes, while a similar transmitarray (TA) approach is suggested in Liu et al. (2021). A circularly polarized beams are produced with the help of antenna array in Li and Zhou (2018). The authors of Liao et al. (2022) have created tunable modes using resonators. In Ming and Shi (2020), adjustable feeding system was incorporated to transmit OAM modes with the help of water-based antenna. It is also important to note that most of these works use PIN diodes for reconfigurable feeds. But in Naseri et al. (2022), the authors proposed a low-profile design that uses varactor diodes to generate tunable OAM modes. This helps to overcome the problem of signal leakage in PIN diodes.

Some antenna structures use separate ports for each OAM mode. For example, in Li et al. (2017a), two different feeding networks are designed. One produces phase shifts of [0, 90, 180, 270] degrees for mode +1, while the other gives [0, −90, 0, −90] degrees for mode −1. In Qin et al. (2019), four OAM modes are generated with four separate antenna feeds. Modes −1 and −2 are created using two different networks placed on the same layer, mode −3 has its own feed on a separate layer, and mode 0 is excited by a central patch antenna. Another method is to use sequential rotation of elements in a circular array, which makes the feeding network simpler. This technique is applied in Zhang and Li (2018) to generate modes +1 and +2 with low-profile feeds. Without this method, the feed design would become more complex and lossy. Radial line slot arrays (RLSA) can also be used to create OAM waves, as shown in Xu et al. (2020), where a 4×4 Butler Matrix is used to generate OAM modes 0, −1, and +1. A quad-mode circular array has been presented in Yoo and Son (2020), where the feeding is done through hybrid couplers linked to ring-shaped transmission lines. Other designs with multiple input ports for different OAM modes are also reported in Bai et al. (2017), Guo and Yang (2016), Liu et al. (2018a), Mazzinghi and Freni (2017), and Palacin et al. (2014).

Metamaterials have been widely used in many areas, such as filter-antenna sensors (Esfandiyari et al., 2022), high-gain all-metal antennas (Lalbakhsh et al., 2021), beam-steerable frequency selective surfaces for phase shifting (Das et al., 2022), and also in UWB and wideband antennas and filters (Adibi et al., 2021; Lalbakhsh et al., 2022). For OAM antennas, several metamaterial-based designs have also been reported (Akram et al., 2019; Huang and Huang, 2022; Li et al., 2017b; Lv et al., 2020; Qin et al., 2018; Wang et al., 2022a,b; Xu et al., 2019; Yang et al., 2019). A complete review of guided-wave metasurfaces for vortex beam generation can be found in Shi et al. (2022). However, most of these metamaterial antennas can only generate one OAM beam, which makes them unsuitable for high data rate communication. To achieve multiple modes, mixed-OAM antennas are often required. Mixed-OAM antennas can work like MIMO systems, sending multiple data streams, but they are very hard to realize with single-port reconfigurable structures. In designs with individual ports, the complexity of feeds becomes a big problem, especially when high gain is needed. To solve this issue, some studies have proposed using TAs and reflectarrays to create mixed-OAM modes. These approaches can provide high gain without complex feeding networks (Cheng et al., 2014; Chung et al., 2022; Feng et al., 2022; Shahmirzadi and Kishk, 2022; Shahmirzadi et al., 2021). For example, Cheng et al. (2014) used perforated SPPs, but they need advanced fabrication. In Shahmirzadi et al. (2021), a two-layer design was used, while Shahmirzadi and Kishk (2022) applied cascaded TAs. In Feng et al. (2022), two off-axis feeds were placed in a reflectarray to combine two OAM modes at the broadside, and Chung et al. (2022) used a similar method but with a TA.

In Naseri et al. (2023), mixed OAM modes of −1 and −2 are created by using a very thin dual-polarized Huygens' metasurface. The design process begins with finding the angular position of each unit cell in a planar TA. The TA must meet two main conditions: first, it should convert the incoming spherical wave into a planar wave; second, it should change it into mixed OAM modes −1 and −2 based on the angular position of the unit cells. The measured results show that the radiation patterns have a null at broadside and clear helical phase fronts. The antenna works at 28 GHz, which is a possible band for 5G networks. The maximum gain achieved is 16 dBi for both X and Y polarizations.

9.7.2 Unit Cell Characteristics

In Naseri et al. (2023), an ultrathin Huygens' metasurface is presented that can provide phase shifts from 0° to 360°. Figure 9.3 shows the layout of the unit cell and its simulation setup in CST software.

The unit cell is printed on both sides of Rogers 4003C, which has a dielectric constant of 3.55 and a thickness of 1.524 mm. Because the cell has a square and symmetric structure, it works in dual polarization. This means that the polarization of the incoming wave does not matter, since both horizontal and vertical parts of the wave interact with the same geometry. In the simulation, the "unit cell" boundary condition is used for the sides of the structure shown in Figure 9.3c. For the top and bottom sides, an open (add space) boundary is applied where the feeds are located. By changing the values of l and g, while keeping their sum equal to 4.4 mm, the required phase compensation from 0° to 360° can be achieved. For the

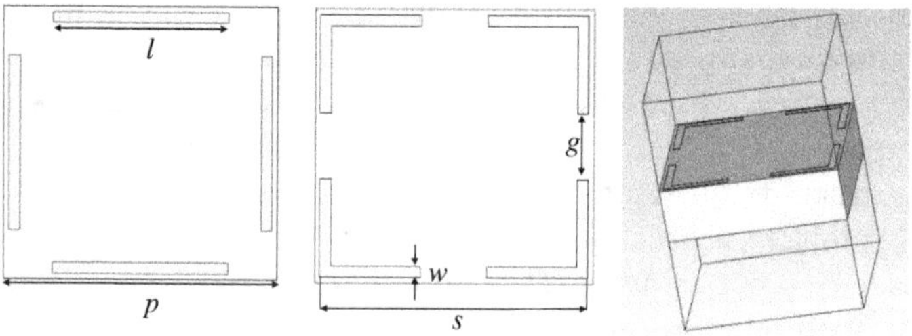

Parameter	Value (mm)
l+g	4.4
p	5
s	4.8
w	0.2

(a) (b) (c) (d)

Figure 9.3 Huygens' metasurface: (a) top view, (b) bottom view, (c) simulation environment, and (d) unit cell parameters and their values. *Source*: Naseri et al. (2023)/ MDPI/CC BY 4.0.

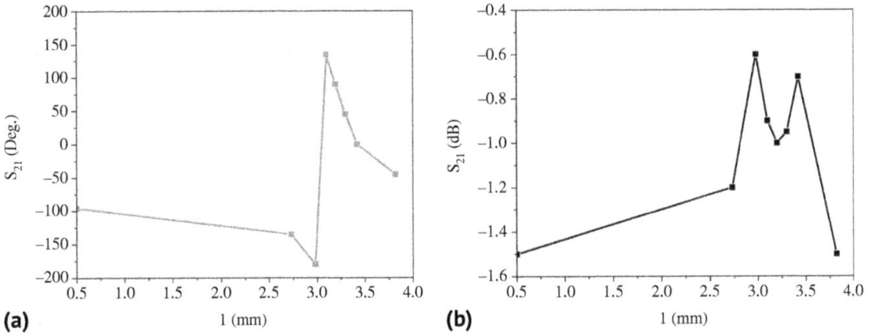

Figure 9.4 Unit cell results for different l values: (a) phase response and (b) amplitude response. *Source*: Naseri et al. (2023)/MDPI/CC BY 4.0.

OAM-mixed TA in this work, only certain quantized phase values are used, which will be explained later. The distance between the unit cell surface and the top/bottom boundary box is chosen as $\lambda_0/2$, and the reference planes are set at the front and back surfaces of the unit cell.

Figure 9.4 shows how the transmitted phase and amplitude change with different values of l. It should be noted that in each step, the condition $l + g = 4.4$ mm is always kept. Looking at Figure 9.4a, the phase response is smooth when l changes from 0.5 to 2.75 mm. After that, a sudden jump appears between 3 and 3.1 mm. This sharp change is only due to the periodic nature of the phase. After this point, the phase continues to decrease until $l = 3.82$ mm. In this study, we only used specific phase values of $0°, 45°, 90°, 135°, 180°, 225°, 270°$, and $315°$. This is because a phase quantization method is applied for generating OAM modes. For the amplitude response shown in Figure 9.4b, the results prove that the Huygens' metasurface unit cell has very low transmission loss, which makes it a good option for producing high-gain OAM waves.

9.7.3 Mixed-OAM generating TA Design

After choosing the right values of l so that the unit cell meets the needed quantized phase values and keeps low transmission loss, the TA can be designed. For this design, two main steps are important. First, the TA must change the received spherical wave into a planar wave. Second, it must create the required mixed OAM modes. These two steps are combined in Equation (9.6) (Yu et al., 2016).

$$\phi_{\mathrm{mn}}^{c} = -\frac{2\pi}{\lambda}\left|\overrightarrow{r_{\mathrm{mn}}} - \overrightarrow{r_f}\right| + \arg\{e^{(j)l_k \cdot \Phi_k}\} \tag{9.6}$$

Here, ϕ^{c}_{mn} is the phase needed for the mn-th element of the TA, which has 400 unit cells (20 × 20). λ is the wavelength in free space at 28 GHz, $\overrightarrow{r_{mn}}$ is the position vector of each unit cell, and $\vec{r_f}$ is the position vector of the feed antenna (horn). Also, l_k is the OAM mode number, and Φ_k is the azimuth angle of each unit cell on the TA plane. To make it clearer, Figure 9.5 shows a simple example of a TA with 6 × 6 elements for illustration.

The first part of Equation (9.6) is related to converting the spherical wave into a planar wave. To achieve a planar wave, all the waves passing through the unit cells of the TA should be in phase. When a horn antenna is used as the feed, the electric fields on the TA plane have different phase values at different points. By using the simulated values of l, it is possible to adjust the phases so that all fields become in-phase.

In the second step, the desired OAM modes are chosen. Using the second part of the equation, the additional phase needed for each unit cell is calculated. Adding these two phase contributions gives the exact phase compensation required to generate a single OAM mode.

To create mixed OAM modes, the central 10 × 10 section of the 20 × 20 TA is assigned to mode −1, while the remaining cells generate mode −2. In other words, the central sub-array produces OAM mode −1, and the outer part of the TA produces mode −2.

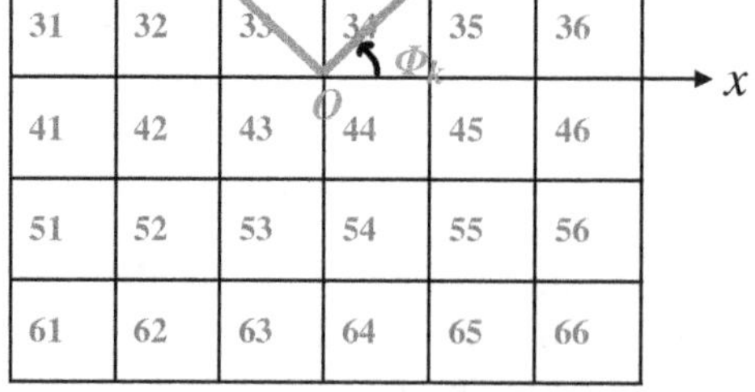

Figure 9.5 Schematic of 6 × 6 elements of TA. *Source:* Naseri et al. (2023)/MDPI/CC BY 4.0.

After calculating all the necessary compensating phases, it is time to utilize quantized values according to Equation (9.7).

$$\phi^c_{\text{mn, quantized}} = \begin{cases} 0 & -22.5 < \phi^c_{\text{mn}} \leq 22.5 \\ 45 & 22.5 < \phi^c_{\text{mn}} \leq 67.5 \\ 90 & 67.5 < \phi^c_{\text{mn}} \leq 112.5 \\ 135 & 112.5 < \phi^c_{\text{mn}} \leq 157.5 \\ 180 & 157.5 < \phi^c_{\text{mn}} \leq 202.5 \\ 225 & 202.5 < \phi^c_{\text{mn}} \leq 247.5 \\ 270 & 247.5 < \phi^c_{\text{mn}} \leq 292.5 \\ 315 & 292.5 < \phi^c_{\text{mn}} \leq 337.5. \end{cases} \tag{9.7}$$

Using the above equation, only eight phase values are considered in the simulation, which helps to reduce the design complexity. In addition, this quantized approach makes it easier to study reconfigurable OAM-generating TAs in the future, using varactor diodes or PIN diodes.

9.7.4 Experimental Verification

The prototype of the designed TA, which has 20×20 unit cells, is fabricated on Rogers 4003C with a thickness of 60 mils, as shown in Figure 9.6a,b. The central 10×10 unit cells are used to generate OAM mode −1, while the remaining cells produce OAM mode −2. In the figure, these two sections are marked with square outlines to show their separation. To verify the generation of mixed OAM modes, the setup shown in Figure 9.6c was used. A probe scanned a square area of $15 \times 15\ \text{cm}^2$, placed 15 cm away from the TA. The TA (antenna under test) was kept fixed, while the probe transmitted signals at each position during the scan. This allowed the TA to detect the amplitude and phase of the transmitted signal at every point. Figure 9.7 shows the resulting phase front and amplitude response. As seen, there is a null at the center of the scanning area, and the two OAM modes (−1 and −2) appear together in the helical phase fronts.

The far-field radiation pattern of the TA is shown in Figure 9.8. A maximum gain of about 16 dBi was measured for both X and Y polarizations, confirming that the TA is dual-polarized. The null at $\Theta = 0$ is also clearly visible.

Mode purity is used to check the quality of the generated OAM modes. For a structure producing a single OAM mode, the purity is theoretically 1. Since the proposed TA generates two mixed OAM modes, the purity depends on the radius of the circle used for calculation. Following Xue et al. (2023), the mode

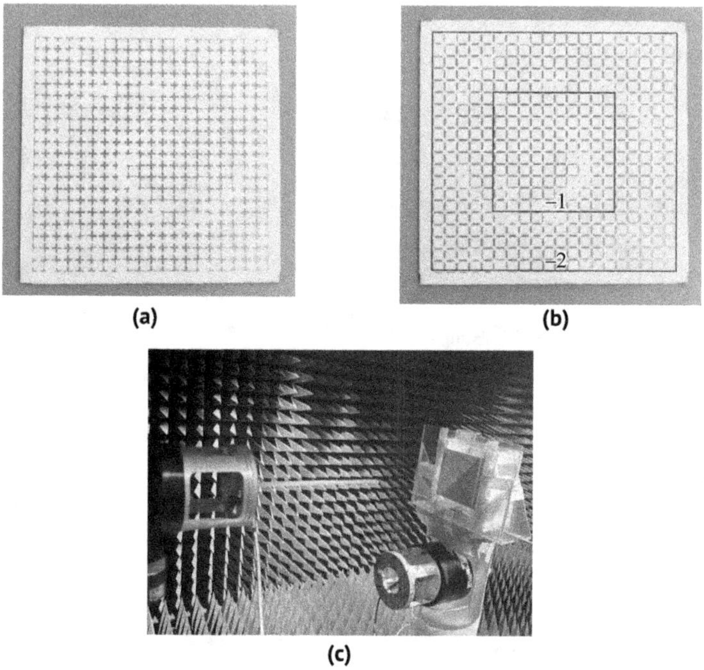

Figure 9.6 Fabricated prototype: (a) top view, (b) bottom view, and (c) near-field measurement setup. *Source*: Naseri et al. (2023)/MDPI/CC BY 4.0.

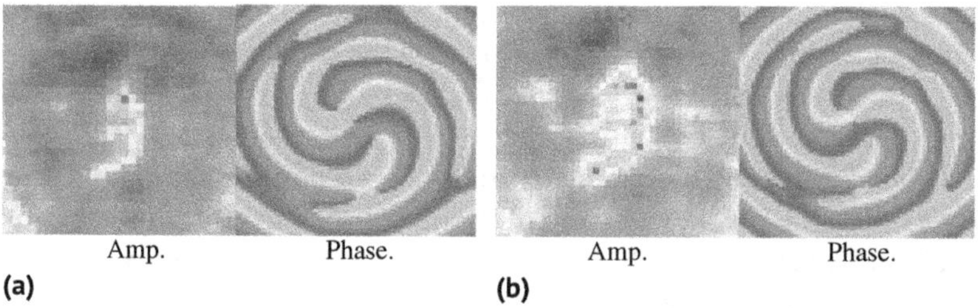

Figure 9.7 Results at 28 GHz: (a) X-polarization and (b) Y-polarization. *Source*: Naseri et al. (2023)/MDPI/CC BY 4.0.

purity was calculated for both X and Y polarizations. For a circle with $r = 4\lambda$, the purity of mode −1 was around 0.65. As the circle size increases, the purity of mode −1 decreases while the purity of mode −2 increases. Some of these results are shown in Figure 9.9. Because the results for X and Y polarizations were very similar, only the X-polarization case is presented. Ideally, the purity should be 0.5 for both modes −1 and −2, but due to fabrication and quantization errors, some impurities are observed.

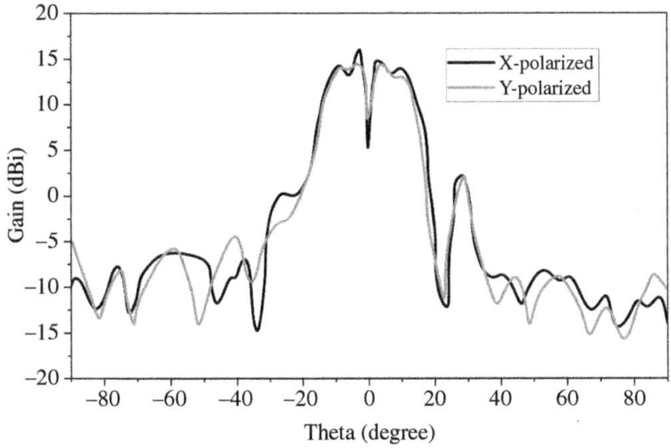

Figure 9.8 X- and Y-polarized measured patterns at 28 GHz. *Source*: Naseri et al. (2023)/ MDPI/CC BY 4.0.

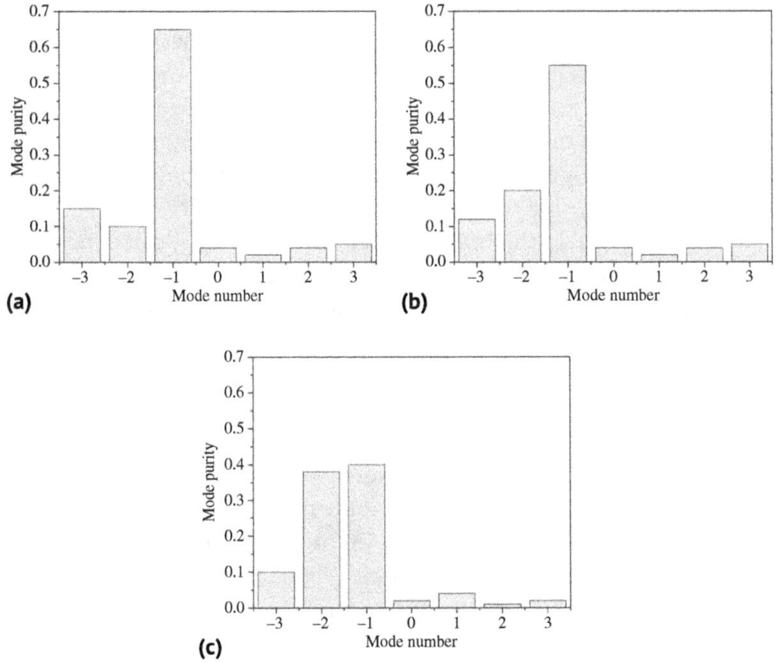

Figure 9.9 Mode purity results: (a) $r = 4\lambda$, (b) $r = 8\lambda$, and (c) $r = 12\lambda$. *Source*: Naseri et al. (2023)/MDPI/CC BY 4.0.

9.7.5 Discussion

As mentioned, OAM antennas can be divided into four main groups. The first group consists of reconfigurable structures that generate one specific mode at a time. The desired OAM mode is selected by controlling PIN diodes or varactor diodes. These antennas are useful as spatial filters. For example, if a mode-tunable antenna is used at the receiver side, it can capture only the signals with the selected mode and ignore unwanted modes.

The second group includes antenna arrays with individual feeding networks. Each feed creates the phase distribution needed for one OAM mode. Since each mode has its own feed, the system becomes more complex. However, such antennas are useful when multiple data streams need to be sent simultaneously at the same frequency. The streams remain separate due to the orthogonality of OAM modes.

The third group includes single antenna elements that generate vortex beams. These antennas usually have low gain and can only produce two OAM modes at a specific frequency, which limits their use in high-data-rate wireless systems.

The fourth group consists of TAs and reflectarrays using different types of metamaterials. These structures usually generate only one OAM mode, which does not increase channel capacity because they cannot send multiple data streams at the same time and frequency. To improve channel capacity, metamaterial-based OAM antennas need to generate mixed OAM modes. By using different OAM modes for different versions of a signal, the chance of receiving a high-quality signal at the receiver is increased. Some works have addressed this (Cheng et al., 2014; Shahmirzadi and Kishk, 2022; Shahmirzadi et al., 2021), but they often face challenges such as fabrication complexity or lack of dual-polarization. Moreover, none of them use quantized phase values during the design, which is a simple and effective way to make the unit cell design easier for TAs.

Besides TAs that generate mixed OAM modes at the broadside, there are some designs that can produce a desired mode at an off-axis position (Chung et al., 2022; Feng et al., 2022). By superimposing phase values on a TA or reflectarray, it is possible to excite two OAM modes at two different locations. According to the reciprocity principle, if horn antennas are placed at the off-axis positions corresponding to the OAM mode directions, the mixed OAM modes will propagate along the on-axis direction. Using this feature, a multiplexing-demultiplexing link can be created with two copies of the structure for transmission and reception.

However, these types of structures cannot generate more than two mixed OAM modes because superimposing the phases on the TA reduces the mode purity. In contrast, in Naseri et al. (2023), the number of mixed OAM modes can be increased easily by adding more sub-TAs, as done in Cheng et al. (2014). Moreover, using two feeding sources to reduce multipath fading makes the system more complex.

The design of Naseri et al. (2023) also has the advantage of being dual-polarized, which is not provided in the previous works.

Regarding the applications of Naseri et al. (2023), it is important to note that it can generate two mixed OAM modes. Since these modes are orthogonal and propagate at the same time, two separate versions of a data stream can be transmitted simultaneously. These streams remain separate, which improves the chance of receiving a high-quality signal. This behavior is similar to MIMO systems, which help reduce multipath fading. Therefore, Naseri et al. (2023) can also provide high-capacity communication.

Another application of mixed OAM modes is in imaging. Using mixed OAM modes from a single aperture allows more information to be captured from the object being imaged, leading to higher-resolution images in next-generation wireless systems.

To summarize, the novelty of Naseri et al. (2023) lies in its polarization independence, thanks to the dual-polarized unit cells. Additionally, using quantized phase values during the design simplifies the structure and reduces design complexity compared to similar works Cheng et al. (2014; Shahmirzadi and Kishk, 2022; Shahmirzadi et al., 2021). This approach also makes it easier to develop mode-tunable OAM TAs by embedding active components such as PIN diodes or Varactor diodes, instead of generating continuous phase values. Moreover, the design proposed by Naseri et al. (2023) is cost-effective because it uses ultra-thin unit cells on a low-cost substrate. In contrast, Cheng et al. (2014) require advanced fabrication equipment, and Shahmirzadi and Kishk (2022) use cascaded substrates. Potential applications of this TA include point-to-point high-data rate communications, high-resolution imaging, and high-quality sensing.

9.7.6 Summary

This study presents a dual-polarized TA capable of generating mixed OAM modes. The design consists of two concentric TAs: the inner 10×10 Huygens' metasurface array generates OAM mode -1, while the outer TA produces mode -2. The OAM modes are created by carefully adjusting the angular position of each unit cell and applying the required phase compensation to convert the spherical wave from the feed into a planar wave. Measured results show helical phase fronts combining modes -1 and -2, with radiation patterns featuring a null at broadside. The maximum gain of 16 dBi is achieved for both X- and Y-polarizations. Mode purity analysis also confirms that the purities of modes -1 and -2 are balanced when considering a larger calculation circle. Overall, this TA design could be a promising solution for increasing channel capacity in next-generation wireless communication systems and for producing high-resolution images in imaging applications.

9.8 Conclusion

OAM antennas at mm-wave frequencies open new dimensions for communication and sensing systems. While significant challenges exist (particularly beam divergence and mode purity), ongoing research in metasurfaces, RIS, and adaptive algorithms is paving the way toward practical deployment. The integration of OAM with 6G paradigms offers the prospect of secure, ultra-high-capacity, and intelligent communication systems.

References

Shima Adibi, M. Amin Honarvar, and Ali Lalbakhsh. Gain enhancement of wideband circularly polarized UWB antenna using FSS. *Radio Science*, 56(1):1–8, 2021.

Muhammad Rizwan Akram, Xudong Bai, Ronghong Jin, Guy AE Vandenbosch, Malin Premaratne, and Weiren Zhu. Photon spin hall effect-based ultra-thin transmissive metasurface for efficient generation of OAM waves. *IEEE Transactions on Antennas and Propagation*, 67(7):4650–4658, 2019.

Xu-Dong Bai, Xian-Ling Liang, Yun-Tao Sun, Peng-Cheng Hu, Yu Yao, Kun Wang, Jun-Ping Geng, and Rong-Hong Jin. Experimental array for generating dual circularly-polarized dual-mode OAM radio beams. *Scientific Reports*, 7(1):40099, 2017.

Li Cheng, Wei Hong, and Zhang-Cheng Hao. Generation of electromagnetic waves with arbitrary orbital angular momentum modes. *Scientific Reports*, 4(1):1–5, 2014.

Hyeongju Chung, Daeik Kim, Eunmi Choi, and Jongwon Lee. E-band metasurface-based orbital angular momentum multiplexing and demultiplexing. *Laser and Photonics Reviews*, 16(6):2100456, 2022.

Priyanka Das, Kaushik Mandal, and Ali Lalbakhsh. Beam-steering of microstrip antenna using single-layer FSS based phase-shifting surface. *International Journal of RF and Microwave Computer-Aided Engineering*, 32(3):e23033, 2022.

Meisam Esfandiyari, Ali Lalbakhsh, Saughar Jarchi, Mohsen Ghaffari-Miab, Hamideh Noori Mahtaj, and Roy BVB Simorangkir. Tunable terahertz filter/antenna-sensor using graphene-based metamaterials. *Materials and Design*, 220:110855, 2022.

Qiang Feng, Xudong Kong, Mingming Shan, Yifeng Lin, Long Li, and Tie Jun Cui. Multi-orbital-angular-momentum-mode vortex wave multiplexing and demultiplexing with shared-aperture reflective metasurfaces. *Physical Review Applied*, 17(3):034017, 2022.

Zhi-Gui Guo and Guo-Min Yang. Radial uniform circular antenna array for dual-mode OAM communication. *IEEE Antennas and Wireless Propagation Letters*, 16:404–407, 2016.

Hui-Fen Huang and Hong-Ming Huang. Millimeter-wave wideband high efficiency circular airy OAM multibeams with multiplexing OAM modes based on transmission metasurfaces. *Progress in Electromagnetics Research*, 173:151–159, 2022.

HuiFen Huang and ZhiPing Zhang. A single fed wideband mode-reconfigurable OAM metasurface cp antenna array with simple feeding scheme. *International Journal of RF and Microwave Computer-Aided Engineering*, 31(2):e22499, 2021.

Xiaonan Hui, Shilie Zheng, Yiling Chen, Yiping Hu, Xiaofeng Jin, Hao Chi, and Xianmin Zhang. Multiplexed millimeter wave communication with dual orbital angular momentum (OAM) mode antennas. *Scientific Reports*, 5(1):10148, 2015.

Le Kang, Hui Li, Jinzhu Zhou, Shufeng Zheng, and Steven Gao. A mode-reconfigurable orbital angular momentum antenna with simplified feeding scheme. *IEEE Transactions on Antennas and Propagation*, 67(7):4866–4871, 2019.

Le Kang, Hui Li, Jinzhu Zhou, and Shufeng Zheng. An OAM-mode reconfigurable array antenna with polarization agility. *IEEE Access*, 8:40445–40452, 2020.

Ali Lalbakhsh, Muhammad U. Afzal, Touseef Hayat, Karu P. Esselle, and Kaushik Mandal. All-metal wideband metasurface for near-field transformation of medium-to-high gain electromagnetic sources. *Scientific Reports*, 11(1):9421, 2021.

Ali Lalbakhsh, Muhammad U. Afzal, Karu P. Esselle, and Stephanie L. Smith. All-metal wideband frequency-selective surface bandpass filter for TE and TM polarizations. *IEEE Transactions on Antennas and Propagation*, 70(4):2790–2800, 2022.

Hui Li, Le Kang, Feng Wei, Yuan-Ming Cai, and Ying-Zeng Yin. A low-profile dual-polarized microstrip antenna array for dual-mode OAM applications. *IEEE Antennas and Wireless Propagation Letters*, 16:3022–3025, 2017a.

Hui Li, Le Kang, and Ke Dong. Generating tunable orbital angular momentum radio beams with dual-circular-polarization and dual-mode characteristics. *IEEE Access*, 8:211248–211254, 2020a.

Long Li and Xiaoxiao Zhou. Mechanically reconfigurable single-arm spiral antenna array for generation of broadband circularly polarized orbital angular momentum vortex waves. *Scientific Reports*, 8(1):5128, 2018.

Weiwen Li, Lei Zhang, Shiyu Yang, Kaiyong Zhuo, Longfang Ye, and Qing Huo Liu. A reconfigurable second-order OAM patch antenna with simple structure. *IEEE Antennas and Wireless Propagation Letters*, 19(9):1531–1535, 2020b.

Yang Li, Xiong Li, Lianwei Chen, Mingbo Pu, Jinjin Jin, Minghui Hong, and Xiangang Luo. Orbital angular momentum multiplexing and demultiplexing by a single metasurface. *Advanced Optical Materials*, 5(2):1600502, 2017b.

Zhen Liao, Yanziyi Che, Leilei Liu, Bai Cao Pan, Ben Geng Cai, Jia Nan Zhou, Guo Qing Luo, and Yongmin Liu. Reconfigurable vector vortex beams using spoof surface plasmon ring resonators. *IEEE Transactions on Antennas and Propagation*, 70(8):6795–6803, 2022.

Baiyang Liu, Guoying Lin, Yuehui Cui, and RongLin Li. An orbital angular momentum (OAM) mode reconfigurable antenna for channel capacity improvement and digital data encoding. *Scientific Reports*, 7(1):9852, 2017.

Baiyang Liu, Sirong Li, Yejun He, Yin Li, and Sai-Wai Wong. Generation of an orbital-angular-momentum-mode-reconfigurable beam by a broadband 1-bit electronically reconfigurable transmitarray. *Physical Review Applied*, 15(4):044035, 2021.

Dandan Liu, Liangqi Gui, Zixiao Zhang, Han Chen, Guochao Song, and Tao Jiang. Multiplexed OAM wave communication with two-OAM-mode antenna systems. *IEEE Access*, 7:4160–4166, 2018a.

Qiang Liu, Zhi Ning Chen, Yuanan Liu, Fuhai Li, Yu Chen, and Zhiguang Mo. Circular polarization and mode reconfigurable wideband orbital angular momentum patch array antenna. *IEEE Transactions on Antennas and Propagation*, 66(4):1796–1804, 2018b.

Huan-Huan Lv, Qiu-Lin Huang, Xiang-Jie Yi, Jian-Qiang Hou, and Xiao-Wei Shi. Low-profile transmitting metasurface using single dielectric substrate for OAM generation. *IEEE Antennas and Wireless Propagation Letters*, 19(5):881–885, 2020.

Agnese Mazzinghi and Angelo Freni. Simultaneous generation of pseudo-Bessel vortex modes with a RLSA. *IEEE Antennas and Wireless Propagation Letters*, 16:1747–1750, 2017.

Jie Ming and Yan Shi. A mode reconfigurable orbital angular momentum water antenna. *IEEE Access*, 8:89152–89160, 2020.

Naseri, Hassan PourMohammadi, Peyman Melouki, Noureddine Iqbal, and Amjad Denidni. Tayeb A. A low-profile antenna system for generating reconfigurable OAM-carrying beams. *IEEE Antennas and Wireless Propagation Letters*, 22(2): 402–406, Early Access. 2022.

Hassan Naseri, Peyman PourMohammadi, Nouredddine Melouki, Fahad Ahmed, Amjad Iqbal, and Tayeb A Denidni. Generation of mixed-OAM-carrying waves using Huygens' metasurface for mm-wave applications. *Sensors*, 23(5):2590, 2023.

Baptiste Palacin, Ksenia Sharshavina, Kevin Nguyen, and Nicolas Capet. An 8×8 butler matrix for generation of waves carrying orbital angular momentum (OAM). In *The 8th European Conference on Antennas and Propagation (EuCAP 2014)*, pages 2814–2818. IEEE, 2014.

Fan Qin, Lulan Wan, Lihong Li, Hailin Zhang, Gao Wei, and Steven Gao. A transmission metasurface for generating OAM beams. *IEEE Antennas and Wireless Propagation Letters*, 17(10):1793–1796, 2018.

Fan Qin, Lihong Li, Yi Liu, Wenchi Cheng, and Hailin Zhang. A four-mode OAM antenna array with equal divergence angle. *IEEE Antennas and Wireless Propagation Letters*, 18(9):1941–1945, 2019.

Arash Valizade Shahmirzadi and Ahmed A. Kishk. Oam carrying vortex beam mode interconversion using modular cascaded transmitarrays. *IEEE Transactions on Microwave Theory and Techniques*, 70(7):3591–3605, 2022.

Arash Valizade Shahmirzadi, Zahra Badamchi, Bahareh Badamchi, and Harish Subbaraman. Generating concentrically embedded spatially divided OAM carrying vortex beams using transmitarrays. *IEEE Transactions on Antennas and Propagation*, 69(12):8436–8448, 2021.

Shi, Hongyu Wang, Luyi Li, Guoqiang Yi, Jianjia Liu, Haiwen Zhang, and Anxue Xu. Zhuo Guided-wave inspired metasurfaces for multifunctional vortex beam generation and manipulation. *Journal of Lightwave Technology*, 41(7):2094–2106. Early Access. 2022.

Heng Wang, Shixing Yu, Na Kou, Zhao Ding, and Zhengping Zhang. Cylindrical holographic impedance metasurface for OAM vortex wave generation. *Applied Physics Letters*, 120(14):143504, 2022a.

Luyi Wang, Hongyu Shi, Jianjia Yi, Liang Dong, Haiwen Liu, Anxue Zhang, and Zhuo Xu. Suspended metasurface for broadband high-efficiency vortex beam generation. *Materials*, 15(3):707, 2022b.

Ziyang Wang, Xiaotian Pan, Fan Yang, Shenheng Xu, and Maokun Li. Real-time mode switching and beam scanning of high-gain OAM waves using a 1-bit reconfigurable reflectarray antenna. *Electronics*, 9(12):2181, 2020.

Jie Wu, Zhongxiang Zhang, Xingang Ren, Zhixiang Huang, and Xianliang Wu. A broadband electronically mode-reconfigurable orbital angular momentum metasurface antenna. *IEEE Antennas and Wireless Propagation Letters*, 18(7):1482–1486, 2019.

Jianchun Xu, Ke Bi, Ru Zhang, Yanan Hao, Chuwen Lan, Klaus D. McDonald-Maier, Xiaojun Zhai, Zidong Zhang, and Shanguo Huang. A small-divergence-angle orbital angular momentum metasurface antenna. *Research*, 2019.

Xin Xu, Agnese Mazzinghi, Angelo Freni, and Jiro Hirokawa. Simultaneous generation of three OAM modes by using a RLSA fed by a waveguide circuit for 60 GHz-band radiative near-field region OAM multiplexing. *IEEE Transactions on Antennas and Propagation*, 69(3):1249–1259, 2020.

Hao Xue, Jiaqi Han, Song Zhang, Yucen Tian, Jianqiang Hou, and Long Li. Co-modulation of spin angular momentum and high-order orbital angular momentum based on anisotropic holographic metasurfaces. *IEEE Transactions on Antennas and Propagation*, 71(5):4594–4599, 2023.

Yan Yan, Guodong Xie, Martin PJ Lavery, Hao Huang, Nisar Ahmed, Changjing Bao, Yongxiong Ren, Yinwen Cao, Long Li, Zhe Zhao, et al. High-capacity millimetre-wave communications with orbital angular momentum multiplexing. *Nature Communications*, 5(1):4876, 2014.

Xianhang Yang, Yonggang Zhou, and Guanteng Wang. A wideband transmission metasurface for generating Bessel beam carrying orbital angular momentum. *International Journal of RF and Microwave Computer-Aided Engineering*, 29(11):e21941, 2019.

Jeong-Ung Yoo and Hae-Won Son. Quad-mode radial uniform circular array antenna for OAM multiplexing. *IET Microwaves, Antennas and Propagation*, 14(8):728–733, 2020.

Shixing Yu, Long Li, Guangming Shi, Cheng Zhu, and Yan Shi. Generating multiple orbital angular momentum vortex beams using a metasurface in radio frequency domain. *Applied Physics Letters*, 108(24):241901, 2016.

Yi-Ming Zhang and Jia-Lin Li. Analyses and full-duplex applications of circularly polarized OAM arrays using sequentially rotated configuration. *IEEE Transactions on Antennas and Propagation*, 66(12):7010–7020, 2018.

10

Conclusion

Millimeter-wave (mm-wave) technology is becoming an important part of modern wireless communication systems. The demand for faster data rates, larger bandwidth, and reliable wireless connectivity continues to increase due to applications such as fifth-generation (5G) and future sixth-generation (6G) networks, autonomous vehicles, high-resolution radar systems, and high-capacity wireless links. These applications require antenna systems that can operate efficiently at very high frequencies while remaining compact and practical for integration into modern electronic platforms.

This book has presented the design and development of antennas operating at mm-wave frequencies. The discussion started with the fundamentals of mm-wave communication systems and the main challenges associated with these high frequencies. Unlike lower frequency systems, mm-wave signals suffer from higher free-space path loss, stronger attenuation through obstacles, and increased sensitivity to environmental conditions. Because of these limitations, antenna design becomes a critical part of system performance.

Several antenna technologies suitable for mm-wave systems were discussed throughout this book. Planar antennas were introduced as one of the most widely used solutions because they are low profile, easy to fabricate, and compatible with printed circuit technologies. Different planar antenna configurations, including microstrip, slot, aperture-coupled, and substrate integrated waveguide (SIW)-based antennas, were explained together with their operating principles and design considerations.

Dielectric resonator antennas (DRAs) were also presented as an efficient alternative for mm-wave applications. Since DRAs do not rely heavily on metallic conductors, they can reduce conductor losses and achieve higher radiation efficiency at high frequencies. Different feeding techniques and material considerations were described to highlight how DRAs can be effectively implemented in practical systems.

The book also examined SIW technology as an important platform for mm-wave antenna design. SIW structures combine the low-loss characteristics of conventional waveguides with the compact size and manufacturability of planar circuits. This makes them highly attractive for integrated antenna systems and antenna arrays used in modern wireless communication platforms.

Wideband antenna techniques were further discussed to address the increasing bandwidth requirements of current wireless systems. Techniques such as stacked structures, parasitic elements, and hybrid configurations were introduced as effective ways to enhance antenna bandwidth while maintaining stable radiation performance.

Antenna arrays were presented as a key component of mm-wave communication systems. Due to the high propagation loss at mm-wave frequencies, antenna arrays are often required to achieve higher gain and directional radiation patterns. Different feeding approaches, including series-fed, parallel-fed, and hybrid arrays, were discussed together with their advantages and practical design considerations.

The book also explored advanced antenna concepts that allow dynamic control of antenna properties. Reconfigurable antennas based on frequency selective surfaces were introduced as a promising approach for adaptive wireless systems. These antennas can adjust their operating characteristics, such as frequency response, polarization, or radiation pattern, which makes them useful for next-generation communication platforms.

Finally, the concept of orbital angular momentum (OAM) antennas was discussed as an emerging research direction. OAM antennas offer the possibility of increasing channel capacity and improving communication security by transmitting multiple orthogonal electromagnetic modes. Although this technology is still developing, it represents an interesting opportunity for future wireless systems.

Overall, this book has shown that designing antennas for mm-wave systems requires a balance between electromagnetic theory, practical design techniques, and system-level understanding. As wireless technologies continue to evolve toward higher frequencies and more complex architectures, antenna design will remain a key area of research and innovation.

Future research will likely focus on several directions. These include the development of compact antenna arrays for mobile and automotive platforms, improved integration of antennas with semiconductor technologies, and the design of intelligent and reconfigurable antenna systems. In addition, emerging concepts such as metasurfaces, reconfigurable intelligent surfaces, and artificial-intelligence-assisted design methods are expected to play an important role in future mm-wave communication systems.

In conclusion, mm-wave antennas will continue to be a fundamental component of next-generation wireless technologies. Continued advances in antenna design, materials, and fabrication techniques will help support the growing need for faster, more reliable, and more efficient wireless communication systems.

Index

Note: Page numbers in *italics* and **bold** refers to figures and tables respectively.

a

active FSS elements
  adaptive FSS-based antenna systems 161–162
  biasing techniques for 151–152
additive manufacturing 24
advanced wireless communication systems 51
AF. *see* array factor (AF)
amplitude modulation (AM) 9
amplitude tapering 113
angular momentum in electromagnetics 171–172
antenna 1, 2, 31
  arrays 194
  importance of 4
  mm-wave. *see* mm-wave antennas
  OAM-based antennas 4
aperture-coupled antennas 3, 40–41, 193
aperture-coupled DRAs 59–60
aperture-type parasitic elements 93–94
array design 21
array factor (AF) 118, 122, 148
  for OAM modes 174
artificial intelligence (AI)
  for adaptive control 164
  AI-assisted design methods 194
  driven optimization for adaptive FSS-based antenna systems 161–162
augmented reality (AR) 10, 13, 15
automotive communications 14
automotive radar 12, 31, 158
  FSS-based reconfigurable antennas 163
  millimeter-wave reconfigurable antennas in 158
  at mm-wave frequencies 108–109, 109
  parallel-fed arrays in 120
  planar arrays in 108–109
  series-fed arrays in 115
  SIW antennas at 43, 77
  systems, DRAs in 62
  traditional 14

b

bandpass FSS 141–142
bandstop FSS 141–142
bandwidth 121, 154
  for antenna arrays 128
  in combination-fed arrays 125
  fractional. *see* fractional bandwidth (FBW)

bandwidth (*Cont.*)
for millimeter-wave planar antennas 48, 128
of mm-wave 2, 14
narrow 36
for single-band SIW antennas 72
wide 85
bandwidth enhancement 21, 40, 87–88
for mm-wave antennas 21
parasitic-element techniques for 90–95
stacked microstrip antenna techniques for 88–90
beam divergence 174–175
beamforming 10, 106
and mobility limitations 19
beam shaping 154
beam steering 106
for combination-fed arrays 124
for mm-wave antennas 21
reconfigurability 144
for SIW 79
biasing techniques for active FSS elements 151–152
binomial taper 113
biomedical imaging, mm-wave reconfigurable antennas at 158
broadband antenna array 178

C

cavity-type single-band SIW antenna 71–72
cellular networks 9
Chebyshev taper 113
chemical etching 31, 34
chip-to-chip wireless links 177
CMOS, integration with compact packaging and 160–161
coaxial probe feeding technique 3, 57–58, *58*
combination-fed arrays 120–121
applications 126
array factor and mathematical model 122
beam steering and pattern control 124
comparison with series and parallel arrays 125, **126**
corporate–series hybrid feed 123
current and phase distribution 122–123
design considerations and equations 123
input impedance 124
loss, bandwidth, and efficiency 125
mutual coupling and matching 125
need for 121–122
parallel-series configuration 123
power division 124
principle of 121
series-parallel configuration 123
typical combination structures 123
compact packaging, integration with CMOS and 160–161
computer simulation technology (CST) 150–151
conductor losses 56–57
connected and autonomous vehicle (CAV) environments 14
conventional waveguides 67
coplanar waveguide (CPW) 3, 21, 51
CPW-fed DRAs 61
CPW-fed microstrip patch antenna 31, *32*, 33, *33*
corporate feeding 78
corporate feed network 119
corporate–series hybrid feed of combination-fed arrays 123
coupling methods 3
coupling-based techniques 81
CST. *see* computer simulation technology (CST)
current distribution of combination-fed arrays 122–123

d

dielectric losses 56–57
dielectric materials for DRAs 54–55
dielectric resonator antennas (DRAs) 3, 5, 51–52, 193
  advantages at mm-wave frequencies 52–54
  applications of 62–63, *62*
  dielectric materials for 54–55
dielectric resonators 21, 174
dielectric substrate 151
digital signal processing (DSP) 176
dipole elements 140, *141*
directivity 147
directors 92
divergence angle 175

e

electromagnetically coupled DRAs 58–59
electromagnetic band gap (EBG) 98
energy harvesting 158, 164
equivalent circuit models 139–140, 150–151
etching
  bulk silicon 23
  chemical 31, 32, 34, 134
  deep reactive ion etching (DRIE) 23
  laser 31, 32, 34, 134
  mm-wave bands fabrication limits affect 146
  printed circuit board (PCB) 23
extended reality (XR) 15

f

fabrication techniques 22, *23*
  and measurement techniques for mm-wave antennas 22
  micromachining and MEMS techniques 23–24
  at mm-wave bands 159–160
  photolithography and PCB etching 23
  simulation and fabrication constraints 22
  3D printing and additive manufacturing 24
far-field scanning 25
FBW. *see* fractional bandwidth (FBW)
feeding techniques 21, 57
  aperture-coupled DRAs 59–60, *60*
  coaxial probe feeding technique 57–58, *58*
  CPW-fed DRAs 61, *61*
  electromagnetically coupled DRAs 58–59, *59*
  for millimeter-wave slot antennas 48
  for mm-wave antennas 21
  for stacked microstrip antenna 90
feed network design for antenna arrays 128
fifth-generation (5G) 9, 31, 51, *62*, 133, 163, 170, 193
  bandwidth and frequency agility 154
  beam steering for 144
  and beyond communication systems 108
  and beyond mobile networks 15
  mm-wave wireless communication systems 156–157
  on-chip and in-package antennas for 48
  SIW-fed 5G antenna *33*
  wireless communication 1, 103
finite element method (FEM) 150
finite integration technique (FIT) 150
flexible substrates 152
fractal-based unit cells 140, *141*
fractional bandwidth (FBW) 154
free-space path loss (FSPL) 12, 157
frequency agility 154
frequency modulation (FM) 9
frequency reconfigurability 142, 162
frequency-reconfigurable antenna 135

frequency selective surfaces (FSS) 4, 6, 135
advanced materials 164
AI and machine learning for adaptive control 164
AI/ML-driven optimization 161–162
applications 156
bandpass 141–142
bandstop 141–142
bandwidth and frequency agility 154
biasing techniques for active FSS elements 151–152
challenges and future trends 159
definition and basic principle of operation 138
design methodologies 149
efficiency and losses at mm-wave 155
energy efficiency and sustainable design 165
equivalent circuit models for 139–140
fabrication and tolerance issues 159–160
fabrication challenges 152
5G and 6G mm-wave wireless communication systems 156–157
FSS-based polarization-reconfigurable antennas 148
fundamentals of 138
future perspective 162
gain enhancement and beam shaping 154
geometry of *145*
geometry of unit cell of absorben *137*
high-frequency (THz) applications 164
hybrid integration 148
integration with CMOS and compact packaging 160–161
integration with mm-wave reconfigurable antennas 146, 153
integration with RIS for smart environments 159
isolation in MIMO configurations 155
key aspects *165*
metasurfaces for multiband and wideband operation 148
with MIMO antennas 149
mm-wave radar and imaging systems 157–158
multi-functional antennas 165
next-generation wireless and sensing systems 163–164
open research directions 164
parametric analysis of unit cell geometry and materials 151
performance metrics 153
with phased arrays 148
polarization-dependent designs 141–142
reconfigurable FSS geometry *136*
as reconfigurable reflector/absorber for beam control 147–148
reflection and transmission coefficient 153
with RIS 149
role in enabling reconfigurability 162–163
role of 135–137
satellite and space communications 157
7×7 array and unit cell *137*
simulation approaches 150–151
as superstrate for gain enhancement 147
as superstrate geometry *136*
thermal and power handling 160
trade-offs between complexity, cost, and performance 155–156
transmission *vs.* reflection mode 138–139

tunable 2D materials for 161
types of unit cells 140–141
ultra-compact integration 164
unit cell design and periodicity considerations 150
wireless power transfer and energy harvesting 158
FSPL. *see* free-space path loss (FSPL)

g
gain enhancement 154
gain reconfigurability 144
geostationary orbit (GEO) 163
Global Positioning System (GPS) 9
graphene 145, 161

h
healthcare and medical imaging, mm-wave frequencies in 14–15
high frequency (HF) 11
high-frequency structure simulator (HFSS) 150–151
high-resolution photomasks 23
high-speed indoor wireless networks 13–14
Huygens' metasurface 180, *180*
hybrid-fed arrays. *see* combination-fed arrays
hybrid techniques 95
advantages of 98–99
design challenges 99
microstrip patch antenna coupled with two parasitic elements *96*
mix of different feeds and radiators 97
mix of slot and patch antennas 96–97
mix of slots and parasitic edges 98
mix of tuning stubs and modified ground 98
mix with electromagnetic band gap 98
practical examples 99

i
IEEE 802. 11ad (WiGig) 13
imaging and sensing systems at mm-wave frequencies 109
imaging systems 133
immersive technologies 15
impedance matching 125
for antenna arrays 128
biasing networks affect 146
CPW-fed DRAs guarantee 61
for mm-wave antennas 21
industrial automation, mm-wave system in 15
Industry 4. 0 15

j
Jerusalem cross 140

l
laser etching 31, 34
left-hand circular polarization (LHCP) 143
linear polarization (LP) 148
line-of-sight (LOS) 10
and blockages, sensitivity to 18
liquid crystal polymer (LCP) 20
liquid crystals 145
lithography 152
low earth orbit (LEO) 109, 157
low frequency (LF) 11

m
machine learning (ML)
for adaptive control 164
driven optimization for adaptive FSS-based antenna systems 161–162
ML-based demodulation 176
machine-to-machine (M2M) communication 15

measurement techniques
  challenges and best practices 26
  for mm-wave antennas 24
  near-field and far-field scanning 25
  radiation pattern and gain measurements 25
  S-parameter measurements using vector network analyzers 24–25
medium frequency (MF) 11
MEMS. *see* micro-electro-mechanical systems (MEMS)
metamaterials 179
metasurfaces 173–174, 194
MF. *see* medium frequency (MF)
micro-electro-mechanical systems (MEMS) 23–24, 106, 136, 144
micromachining 23–24
microstrip antennas 34
  advantages at mm-wave frequencies 35
  applications 37
  basic structure and operation 34–35
  design considerations and challenges 36, *36*
  design evolution *38*
  design example 37–40
  rectangular microstrip patch antenna *34*
  S_{11} of design evolution *39*
microwave band. *see* microwave region
microwave region 12
millimeter-wave (mm-wave) 67, 193
  challenges at mm-wave bands 146
  frequency band 1
  millimeter-wave systems 86
  radar and imaging systems 157–158
millimeter-wave dielectric resonator antennas 51
  advantages of 52–54
  applications of DRAs 62–63
  conductor and dielectric losses 56–57
  dielectric materials for DRAs 54–55
  dielectric resonator antennas (DRAs) 51–52
  dominant resonant modes *52*
  feeding techniques 57–61
  key material considerations for *56*
millimeter-wave planar antenna arrays
  amplitude distribution 127
  applications of planar arrays 108–109
  bandwidth 128
  challenges in 109–110
  combination-fed arrays 120–126
  design considerations for 107–108, *107*, 127
  element spacing 127
  feed network design 128
  impedance matching 128
  importance of 106
  mutual coupling 128
  parallel-fed arrays 115–120
  phase control 127
  physical and fabrication constraints 128
  planar antenna arrays 103–104
  series-fed antenna arrays 110–115
  2×2 planar patch array *104–105*
millimeter-wave planar antennas
  aperture-coupled antennas 40–41
  applications of 33–34
  coplanar waveguides (CPW)-fed microstrip patch antenna 31, *32*, 33
  design considerations and challenges 32
  microstrip antennas 34–40
  reason for using planar antennas for mm-wave 32
  SIW antennas 41–46
millimeter-wave reconfigurable antennas 142
  challenges at mm-wave bands 146
  design methodologies 149–153
  frequency reconfigurability 142

gain and beam steering reconfigurability 144
integration of FSS with 146–149
overview of reconfigurability 142
performance metrics 153–156
polarization reconfigurability 142
radiation pattern reconfigurability 142–144
techniques for reconfiguration 144–145
millimeter-wave slot antennas 46–49
applications and advantages 48
configuration of mm-wave SIW-based slot antenna *47*
design note for arrays 49
feeding mechanisms 48
principle of operation 47–48
slot width and bandwidth considerations 48
millimeter-wave wireless communication systems 9–10
applications of mm-wave frequencies 15
automotive and V2X communications 14
beamforming and mobility limitations 19
challenges in antenna and circuit design 18–19
design strategies for mm-wave antennas 20–22
fabrication and measurement techniques for mm-wave antennas 22–26
5G and beyond mobile networks 15
healthcare and medical imaging 14–15
high free-space path loss 16–17
high-speed indoor wireless networks 13–14
immersive technologies 15
impairments in 16
industrial automation and smart factories 15
microwave region 12
millimeter-wave 12
poor penetration through obstacles 17
rain, fog, and atmospheric absorption 18
RF spectrum 11, *11*
security screening and surveillance 15–16
sensitivity to line-of-sight and blockages 18
thermal and noise issues 19–20
THz band 12–13
very low to high frequencies 11
VHF and UHF bands 12
wireless backhaul and fronthaul links 14
MIMO. *see* multiple-input multiple-output (MIMO)
mixed-OAM antennas 179
mixed-OAM generating TA design 181–183, *182*
MMICs. *see* monolithic microwave integrated circuits (MMICs)
mm-wave. *see* millimeter-wave (mm-wave)
mm-wave antennas
array design and beam steering 21
challenges of antenna design at mm-wave frequencies 134
design strategies for 20, *20*
fabrication and measurement techniques for 22–26
feeding techniques 21
impedance matching and bandwidth enhancement 21
integration with packaging 21–22
material considerations 20
simulation and fabrication constraints 22
size and miniaturization 21

molybdenum disulfide (MoS_2) 161
monolithic microwave integrated circuits (MMICs) 48
multiband SIW antennas 72. *see also* single-band SIW antennas
  advantages and application context 77
  concept of multiband operation 73
  feed structures and isolation between bands 77
  placement of vias and multiband stability 76–77
  self-diplexing SIW antennas 73–74
  self-quadruplexing SIW antennas 75–76
  self-triplexing SIW antennas 74–75, *75*
multiband SIW antennas 72–77
multi-functional antennas 165
multilayer PCB designs 161
multipath propagation 176
multipath sensitivity of OAM antennas 176
multiple-input multiple-output (MIMO) 10, 86
  antennas 149
  isolation in 155
  transceivers 106
multi-resonant structures 21
mutual coupling
  for antenna arrays 128
  and losses at parallel-fed arrays 119–120
  and matching for combination-fed arrays 125
  for SIW 79–80

n
near-field scanning 25
next-generation wireless and sensing systems 163–164
nonline-of-sight (NLOS) communication 18, 159

o
OAM. *see* orbital angular momentum (OAM)
on-chip feeding mechanisms 21
orbital angular momentum (OAM) 4, 6, 169
  angular momentum in electromagnetics 171–172
  antennas 170, 171, 194
  applications of antennas 176
  array factor for 174
  beam divergence 174–175
  challenges in antennas 174, *175*
  chip-to-chip wireless links 177
  concept of 170
  design example 178
  dielectric resonator modes 174
  experimental verification 183, *184*, *185*
  fundamentals of 171
  future directions for 177
  generation 172, *173*
  groups of 186–187
  literature review 178–180
  metasurfaces 173–174
  mixed-OAM generating TA design 181–183, *182*
  mode purity and cross-talk 175
  multipath sensitivity 176
  orthogonality of OAM modes 172
  phased arrays with progressive phase shift 173
  point-to-point high-capacity links 176
  radar and imaging applications 177
  receiver architectures 176
  secure communications 177
  spiral phase plates 172
  uniform circular array (UCA) 173
  unit cell characteristics 180–181
orthogonality of OAM modes 172, *175*, 176, 186

p

parallel-fed arrays 115, *116–117*. *see also* series-fed antenna arrays
applications of 120
array factor and radiation pattern 118
corporate feed network 119
mutual coupling and losses 119–120
parallel feeding basics 116–117
power division and phase control 118–119
tree or radial feed network 119
types of 119
parallel-series configuration of combination-fed arrays 123
parametric analysis of unit cell geometry and materials 151
parasitic-element techniques
aperture-type parasitic elements 93–94
bandwidth and gain trade-off 94
for bandwidth enhancement 90
basic idea of 91
benefits for mm-wave systems 94
different parasitic-element positions 93
directors and reflectors 92
microstrip patch antenna coupled with two parasitic elements *91*
multiple parasitic elements 92–93, *93*
practical design challenges 95
printed parasitic structures 93
spacing between radiator and parasitic element 91–92
patch-based SIW arrays 79
patch elements 140
patch-type FSS 139
pattern control
for combination-fed arrays 124
for SIW 79
PBC. *see* periodic boundary condition (PBC)
PCR. *see* polarization conversion ratio (PCR)
periodic boundary condition (PBC) 150
periodicity of FSS 138, 150, 152
periodic spacing 151
phase-change materials 145
phase control at parallel-fed arrays 118–119
phase control for antenna arrays 127
phased arrays
FSS with 148
with progressive phase shift 173, *173*
phase distribution of combination-fed arrays 122–123
photolithography 23, 34
PIN diodes 144
planar antenna arrays 103–104
applications of 108–109
design considerations for mm-wave planar arrays 107
importance of mm-wave planar arrays 106
2×2 planar patch array *104–105*
planar antennas 3, 32, 193
point-to-point high-capacity links 176–177
polarization conversion ratio (PCR) 148
polarization-dependent FSS 141–142
polarization reconfigurability 142, 162
polarization-reconfigurable antenna 135
power division
of combination-fed arrays 124
at parallel-fed arrays 118–119
printed circuit board (PCB) etching 23, 31
printed parasitic structures 93
progressive phase shift, phased arrays with 173
proximity coupling 33

q

Q-factor (*Q*) 91

r

radar and imaging applications
  automotive radar 158
  biomedical imaging 158
  of millimeter-wave slot antennas 48
  mm-wave radar and imaging systems 157
  mm-wave reconfigurable antennas with FSS at 163
  of OAM antennas 177
  security and surveillance 157
radial feed network 119
radiation efficiency 155
radiation pattern 118, 144
  analysis 114
  and gain measurements 25
  reconfigurability 142–144
radio frequency (RF) 5, 31, 106, 133
reconfigurability 142
  frequency 142
  FSS role in enabling 162–163
  gain and beam steering 144
  polarization 142
  radiation pattern 142–144
reconfigurable antennas 134–135, 194
reconfigurable FSS
  beam shaping and 154
  geometry *136*
  integration with smart environments 164
  thermal and power handling constraints of 160
  tunable 2D materials for 161
reconfigurable intelligent surfaces (RIS) 149, 170, 194
  FSS with 149
  integration with RIS for smart environments 159
reflection coefficient 147, 153
reflection mode FSS 138–139
reflectors 92
relative permittivity of dielectric materials 54, 55
resonance frequency of unit cell 150
RF. *see* radio frequency (RF)
RHCP. *see* right-hand circular polarization (RHCP)
right-hand circular polarization (RHCP) 143
RIS. *see* reconfigurable intelligent surfaces (RIS)
Rogers RT/duroid 20

s

satellite and point-to-point links, DRAs in 63
satellite and space communication 9, 133, 143
  at mm-wave frequencies 109
  mm-wave reconfigurable antennas with FSS at 157, 163
satellite links 31
security screening and surveillance, mm-wave system in 15–16
self-diplexing SIW antennas 73–74
self-quadruplexing SIW antennas 75–76
self-triplexing SIW antennas 74–75, *75*
series-fed antenna arrays 110. *see also* parallel-fed arrays
  advantages of 114
  array factor of series-fed arrays 112–113
  basic concept of 111–112, *111*
  design of series-fed arrays 113
  limitations 114–115
  practical applications 115
  radiation pattern analysis 114
  series-fed microstrip array 115
  transmission line model 114
series feeding 78
series-parallel configuration of combination-fed arrays 123

Shannon capacity formula 156
Shannon–Hartley theorem 19
Shannon's theorem 10
signal-to-noise ratio (SNR) 10
silicon-based substrates 20
simple LC resonator model 139
simulation approaches 150
single-band SIW antennas 69. *see also* multiband SIW antennas
  cavity-type single-band SIW antenna 71–72
  design process summary 72
  radiation and bandwidth considerations 72
  rules for via placement 70–71
  SIW-DRA-based single-band antenna *70*
  waveguide dimensions and effective width 69–70
sixth generation (6G) 31, 51, *62*, 133, 163, 169, 193
  bandwidth and frequency agility 154
  combination-fed arrays used in 126
  development of 13
  mm-wave antennas for 162
  in mm-wave communication systems 37
  mm-wave wireless communication systems 156–157
  on-chip and in-package antennas for 48
  planar mm-wave antennas in 33
  reconfigurable intelligent surfaces as major part of 159
  wireless communication 1
slot antennas 193
slot arrays 79
slot elements 140
slot loading 21
slot-type FSS 139
smart environments
  integration with RIS for 159
  mm-wave reconfigurable antennas with FSS at 164
smart factories, mm-wave system in 15
SNR. *see* signal-to-noise ratio (SNR)
SOLT. *see* standards like short-open-load-through (SOLT)
S-parameter measurements using vector network analyzers 24–25
spiral phase plates 172
stacked microstrip antenna techniques
  for bandwidth enhancement 88
  basic structure and bandwidth problem 89
  coupling distance and material effects 89
  feeding techniques 90
  mechanical and manufacturing considerations 90
  principle of bandwidth enhancement 89
standards like short-open-load-through (SOLT) 24
steering angle 144
sub-6 GHz spectrum 9
substrate integrated waveguide based mm-wave antennas 67
  multiband SIW antennas 72–77
  single-band SIW antennas 69–72
  SIW antennas arrays 77–80
  two-port SIW structure *68*
  wideband SIW antennas 80–81
substrate integrated waveguides (SIWs) antennas 3, 5, 21, 33, 41–42, 67, 80, 193, 194
  antennas arrays 77, *78*
  applications of 43–44
  array antennas 43
  array configurations 78
  beam steering and pattern control 79
  cavity-backed patch antennas 42
  design considerations 43
  design example 44–46

substrate integrated waveguides (SIWs)
antennas (*Cont.*)
for mm-wave 42
mutual coupling considerations 79–80
performance at mm-wave frequencies 80
reflection coefficient of *45–46*
SIW-based microwave and mm-wave antenna configuration *44*
SIW-fed 5G antenna *33*
slot antennas 43
tapered SIW horn antennas 43
types 42
system-in-substrate (SiS) applications 68

t
tapered SIW horn antennas 43
television broadcasting 9
terahertz (THz) 4, 11, *11*, 162
band 12–13, 145
in FSS design 164
for ultra-broadband communication 177
thermal and power handling constraints of reconfigurable FSS 160
3D printing 24, 152
through-silicon vias (TSVs) 161
thru-reflect-line (TRL) 24
THz. *see* terahertz (THz)
tolerance issues at mm-wave bands 159–160
trade-offs
between complexity, cost, and performance 155–156
optimization in aperture-coupled antennas 41
traditional sub-6 GHz frequency bands 13
traditional wireless communication systems 9
transmission coefficient 153
transmission line *60*, 67
conductor loss for 18
CPW 61
important for antenna matching 87
losses 120, 134
model in series-fed arrays 114
parallel-fed arrays 115
planar 67
SIW for 68
transmission mode FSS 138–139
transmitarray (TA) approach 178
transverse electric (TE) modes 41
tree feed network 119
triplexing SIW antenna 74
TRL. *see* thru-reflect-line (TRL)
TSVs. *see* through-silicon vias (TSVs)
tunable 2D materials for FSS-based reconfigurable antennas 161
tunable FSS 143, 145, 149
two-dimensional (2D) topology 31

u
ultra-high-definition (UHD) 10
ultra high frequency (UHF) 12
uniform circular array (UCA) 173
uniform taper 113
unit cell
characteristics of OAM antennas 180–181
design 150
geometry and materials, parametric analysis of 151
types of 140–141

v
vanadium dioxide (VO_2) 161
varactor diodes 144
vector network analyzers (VNA) 24
S-parameter measurements using 24–25
vehicle-to-everything (V2X) communications 14
very high frequency (VHF) 12

very low frequency (VLF) 11
virtual reality (VR) 13, 15

W

wideband antenna techniques 194
wideband millimeter-wave antennas
  bandwidth enhancement techniques 87–88
  hybrid techniques 95–99
  parasitic-element techniques for bandwidth enhancement 90–95
  stacked microstrip antenna techniques for bandwidth enhancement 88–90
  usage of 85–87
wideband SIW antennas 80–81. *see also* multiband SIW antennas; single-band SIW antennas
wide bandwidth 85
width of the patch (*W*) 35
Wi-Fi. *see* wireless fidelity (Wi-Fi)
wireless backhaul and fronthaul links 14
wireless communication systems 1, 31, 133, 169. *see also* millimeter-wave wireless communication systems
  advanced 51
  DRAs in 62
  traditional 9
wireless fidelity (Wi-Fi) 9
wireless local area networks (WLANs) 9
wireless power transfer (WPT) 158
  mm-wave reconfigurable antennas with FSS at 163
wireless sensing and imaging, DRAs in 63
WLANs. *see* wireless local area networks (WLANs)
WPT. *see* wireless power transfer (WPT)

X

XR. *see* extended reality (XR)

www.ingramcontent.com/pod-product-compliance
Lightning Source LLC
LaVergne TN
LVHW011440121026
849957LV00003B/25
9781394370214